Human Reproduction and Development

P9-BIN-374

Saunders Modern Biology Series

Published Titles

Human Reproduction and Development

Casimer T. Grabowski

University of Miami

SAUNDERS COLLEGE PUBLISHING

Philadelphia   New York   Chicago
San Francisco   Montreal   Toronto
London   Sydney   Tokyo   Mexico City
Rio de Janeiro   Madrid

Address orders to:
383 Madison Avenue
New York, NY 10017

Address editorial correspondence to:
West Washington Square
Philadelphia, PA 19105

Text Typeface: Souvenir
Compositor: The Clarinda Company
Acquisitions Editor: Michael Brown
Project Editor: Janis Moore
Copy Editor: Bonnie Boehme
Managing Editor and Art Director: Richard L. Moore
Design Assistant: Virginia A. Bollard
Text Design: Caliber Design Planning, Inc.
Cover Design: Caliber Design Planning, Inc.
Text Artwork: Vantage Art, Inc.
Production Manager: Tim Frelick
Assistant Production Manager: Maureen Read

Cover: Scanning electron micrograph of the sperm of a sea urchin *(Arbacia punctulata)* penetrating the egg. (Courtesy of Dr. Don W. Fawcett and Dr. Everett Anderson.)

Library of Congress Cataloging in Publication Data

Grabowski, Casimer T.
    Human reproduction and development.

    Bibliography: p.
    Includes index.

    1. Human   reproduction.   I. Title.   [DNLM:
    1. Reproduction. 2. Embryo. WQ 205 G728h]
QP251.G8   1983   612'.6   82-60501

ISBN 0-03-061534-8

HUMAN REPRODUCTION AND DEVELOPMENT   ISBN 0-03-061534-8

© 1983 by CBS College Publishing. All rights reserved. Printed in the United States of America.
Library of Congress catalog card number 82-60501.

3456   090   987654321

35,863

CBS COLLEGE PUBLISHING
Saunders College Publishing
Holt, Rinehart and Winston
The Dryden Press

QP
251
.G8
1983

Grabowski, Casimer
T.

    Human reproduc-
tion and develop-
ment

To
Doris

_____

CAMROSE LUTHERAN COLLEGE
LIBRARY

CAPITAL LUTHERAN COLLEGE
LIBRARY

Preface

The primary aim of this textbook is to present an elementary but comprehensive survey of the many biological aspects of reproduction for students with little or no background in biology. I have made no assumptions other than that the student knows what a cell is and has some rudimentary concept of bodily functions. A working vocabulary of words and concepts is carefully developed as the text progresses. All significant technical terms are printed in **boldface** the first time they are introduced. These and other words are also defined in an extensive glossary.

This book is based on the lectures I give in a course of the same name for nonscience majors that I have taught at Miami and Nassau. The student population has varied in these classes from freshman to senior to postgraduate and from business, music, and education majors to psychology, sociology, and pre-nursing students. Though it is not an easy course, most students find it more interesting than they expected, and very few drop out. In addition to nonscience majors, paramedical students and those wishing to read on their own will find the book useful.

The book is divided into three parts. Unit One, Anatomy and Physiology of the Reproductive System, presents in moderate depth the structure and function of the human reproductive organs. Basic concepts of cell biology, organ architecture, and physiological control mechanisms are introduced and applied to the study of the reproductive organs. The complexities of the female cycle, in particular, are gradually but thoroughly developed. The emphasis both in this section and throughout the text is on normal structure and function, but discussions of common problems and disorders, particularly those that young people are likely to encounter, are also included.

Unit Two, Development of the Human Body, is a unique feature of this book. Most presentations of human development fall into two categories. The first is a relatively superficial type, dealing mainly with the way the embryo looks at two weeks, two months, and so on. At the other extreme are the textbooks of development used in advanced college courses and in medical schools. There is very little in between. I have bridged this gap by presenting an elementary description of the way the body is sculpted from simple rudiments and how adult relationships of organs are achieved. I have also incorporated information on the dynamics of development and on the mechanisms that shape and control the complex process of development.

The basic knowledge presented in the first two units is applied to some more pragmatic aspects of reproductive biology in Unit Three. The essential features of human heredity are presented along with a discussion of genetic problems and what can be done about them. Another chapter is devoted to a subject of considerable current interest—birth defects, the agents that can induce them, the research that is helping to delineate the problem, and a rational approach to prevention. The various phases of family planning are discussed in detail, again utilizing the basic background that has been acquired. A chapter on sexuality, emphasizing psychosocial aspects, is included for perspective. The final chapter looks at probable future developments in reproductive biology.

The emphasis in this book is on basic biology as it relates to reproduction. Only a limited amount of material on the psychological, sociological, and ethical aspects of sexuality and reproduction is included, but I have not ignored these considerations when writing any of the chapters. Many of the topics were chosen because of their relevance to these areas of concern. Insofar as possible, I have tried to maintain throughout an objective, nonjudgmental presentation of facts, for I believe that this is the best preparation for making ethical personal decisions in this area.

I wish to acknowledge the assistance of the many individuals who helped me make this book a reality: my wife, Doris, who consulted, edited, typed, and provided generous doses of encouragement; my colleagues Bruce Grayson and John Rogers, who edited the manuscript; reviewers whose editorial comments were very valuable, including J. L. Hart (George Mason University), H. Duane Heath (California State University at Hayward), and H. Rauch (University of Massachusetts); my typists Marysse Lobean, Jennie Myers, and Jorge Lopez; the publishers and authors who have allowed me to use their illustrations; and the editors at Saunders College Publishing, especially Michael Brown, Biology Editor, for his persistence and encouragement, and Project Editor Janis Moore and her staff. I would also like to thank the many students I have had in "Human Reproduction and Development" for their enthusiasm and the feedback that helped me develop the format for these chapters.

C. T. GRABOWSKI
Miami, Florida
January 1982

Contents

20   Human Reproductive Biology in the Future   226

Glossary of Technical Terms   233

Selected Bibliography   251

Index   255

CHAPTER 1

Introduction: Reproductive Biology Yesterday and Today

Prehistoric Attitudes Toward Reproduction

Presumably you are reading this book because you are interested in reproductive biology. The current interest in and open discussion of sex and sexual problems are relatively new aspects of our contemporary Western culture, but the enthusiasm this generation displays is not a new phenomenon. The earliest evidence we have of human thought in prehistoric times shows that early people were strongly interested in the reproduction of the human species as well as that of animals and plants. Their interest did not stem merely from a desire for sexual gratification; reproduction was important for the survival of the family and the tribe. Reproduction of the animals and plants they ate was also important for their survival. The very earliest human artifacts that have been found clearly indicate that early people were not simply interested in reproduction but, in the opinion of prehistory experts, were also obsessed with animal and human fertility.

The fertility figurines they left behind were of several types. Drawings and sculptures of copulating and pregnant animals were not uncommon (Fig. 1–1). The earliest known carvings of human figures are females with enlarged breasts, abdomens, and genitalia. These are the so-called Venuses of Europe, carved about 20,000 years ago and found in caves from France to Siberia (Fig. 1–2). Some were fashioned from limestone, others from mammoth tusks. From their appearance, it is apparent that these figures were undoubtedly created for ritual or magic purposes associated with fertility rites. This conclusion is supported by the fact that they were often created in deep underground grottoes, which could have been reached by primitive people only with arduous effort and significant peril.

Figure 1–1 The mating bison of the cave of Le Tuc d'Audoubert. The figures, molded from clay, were found in a high-ceilinged grotto 2 treacherous miles below the ground. Ca 20,000 BC. (Photo by Achille Weider, Zurich, Switzerland.)

Although the history of people in the New World does not date back as far as in the Old World, it is interesting to note that the oldest sculptures found in the Americas are obviously fertility figures. These are the Venuses of Valdivia, from coastal Ecuador, and they date back to approximately 7000 years BC. (Fig. 1–3). The typical Venus is sculptured of ceramic material and ranges from 3 to 6 inches in height. The breasts and genitalia are emphasized, but the faces, arms, and legs are relatively insignificant. These Venuses appear to have been gently broken, as if in ceremonial fashion.

Another type of figurine common in many prehistoric cultures is the ithyphallic male figure (with a huge penis), which obviously symbolizes male fertility magic (Fig. 1–4). Disembodied phalluses and phalluses incorporated into ceremonial urns and maces are also very common. Other figurines and draw-

Figure 1–2 The Venus of Lausanne. She is holding a ram's horn in a seemingly ceremonial fashion. Ca 20,000 years BC. (From Bazin: *The History of World Sculpture*. Lamplight Publishing, Weert, Netherlands, 1968.)

Figure 1–3 Two Venuses of Valdivia, one of them pregnant. These ceramic figurines were crafted about 5000 BC.

ings from prehistoric cultures all over the world illustrate pregnancy, labor, and intercourse.

You might think these sculptures and drawings are examples of primitive pornography, but remember that the concept of pornography is really an invention of civilized people. Reproduction was very important to primitive people, and these artifacts are generally believed to be evidence of the deep significance they attached to the process. Sex and reproduction were obviously associated intimately with our ancestors' concepts of life, death, religion, and a mystical relationship to "Mother Earth." The preponderance of such illustrations and figurines with obvious reference to fertility unquestionably illustrates

Figure 1–4 Ithyphallic figure from prehistoric Peru, a hollow vessel probably used for dispensing a fertility potion. (From Kaufmann-Doig: *Sexual Behaviour in Ancient Peru*. Kompaktos, Lima, Peru, 1979.)

primitive attempts to influence the process by magical means. Reproduction was a unique phase of biology that, unlike physiology or ecology, could not be ignored. It was a vitally important process but also a mysterious process poorly understood and hence associated with much magic and superstition.

Historical Concepts of Development

The formal recorded history of thought on the nature of development (the hows and whys of the process) started with the ancient Greek philosopher-scientists, as did many ideas of our Western civilizations. As early as 420 BC, Hippocrates looked at chick embryos (individuals at an early stage of development) and made some deductions from these observations. Aristotle personally observed embryonic development in many species of animals, including the human. In 384 BC he published these observations and his ideas about development in a book, *On the Generation of Animals,* which includes a remarkably accurate classification of animals based on how they reproduce and develop. Aristotle's main idea about the nature of development was that it was a process in which form and structure developed from formless material, a process of **epigenesis** (new formation). He further believed that the female supplied the inert or nutritive material, whereas the semen supplied the spark, the "vital principle," for development. He felt that the menstrual blood, or its equivalent in species that do not menstruate, was used for nutrition, since there is no menstruation during development. The seminal essence acted on this nutritive material and gave it form and organization as the embryo developed from a simple to a complex entity. Aristotle, of course, was the infallible scientific authority in Western culture through the Middle Ages, and consequently very little was added to the history of embryology until the late seventeenth century.

The concept of **preformation** was born at this time. It was largely promoted by Swammerdam in the 1670s. Swammerdam was a fine biologist who studied, among other things, butterfly metamorphosis. It appeared to him that within the chrysalis, or pupal case, there was a neatly folded butterfly. He thought that he could also visualize this same butterfly neatly folded within the caterpillar that preceded the pupa and within the egg itself. He developed his idea of encasement and concluded that within the egg was the organization of the adult, within the egg of the adult was the organization of an individual of the next generation, and so on. This concept was very stimulating to Renaissance scientists, philosophers, and theologians, very often one and the same person. It was simpler to understand than the complexities of epigenetic development. Preformation was a theory that said that everything had been worked out at the time of Creation and that the size of the individual was the only thing that changed from generation to generation. This concept was explored to its fullest. It was speculated that the tiny preformed infants of all

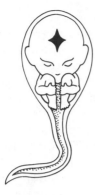

Figure 1–5 A 1694 illustration of a human figure, a homunculus, inside the head of a sperm. (From Katchadourian and Lunde: *Fundamentals of Human Sexuality*, 3rd ed. Holt, Rinehart and Winston, New York, 1980.)

future generations of humans were present within the ovaries of Mother Eve. Some believed that there would have to be a limit to the number of these preformed infants, or homunculi, in Mother Eve's ovaries and that the human race would expire when this supply was exhausted. These ideas were reinforced by discoveries of the mammalian ovum (actually the follicle) by Graaf in 1672 and of the sperm by Leeuwenhoek in 1677.

A new twist to these ideas was added by Hamm, a student of Leeuwenhoek's, who, when he observed sperm cells, thought that he could actually see homunculi in them (Fig. 1–5). He could see tiny preformed frogs in frog sperm, preformed rats in rat sperm, and so forth, as well as little humans in human sperm. He believed that the idea of preformation was right but thought that the preformed individual was in the male rather than the female line.

That was the status of the problem until the 1760s, when the pendulum began to swing back toward the concept of epigenesis. A student named Kaspar Friedrich Wolfe very critically examined the earliest stages of chick development and saw for the first time that (1) the intestine, a delicate midline tube in the young embryo, formed by the folding of a flat layer of cells and (2) blood vessels formed anew from undifferentiated tissues and in areas where previously there were none. The concept of epigenetic development became more accepted from then on as more and more careful observations were made, even though it was vastly more difficult to comprehend than the concept of preformation.

The Scope of Reproductive Biology and This Book

Development of a new individual of the next generation is the ultimate product of reproduction. This phase of reproductive biology is covered in detail in Unit Two of this book. Development, that is, a successful pregnancy, is a rigorous test of the adequacy of the complete reproductive systems of two individuals. This magnificent process can occur only in a suitable incubator in which the developing human is nurtured with a complex and constantly changing bath of nutrients and hormones. Before this can happen, a mature egg must be fertilized by an adequate sperm. Before the egg and the sperm can be brought into proximity, the couple must have functional reproductive organs as well as the ability and desire to use them. The structure and function of the reproductive system are discussed in Unit One, along with the biology of other parts of the body that are of direct importance to reproduction and some common problems that can affect this system.

The facts of basic biology acquired in the first two units are utilized to explore some of the more applied phases of reproduction. The extent to which the unborn child is affected by heredity, environment, and the food and drugs its parents take; birth control and other parameters of human fertility; and the scientific innovations of the near future that can change your reproductive outlooks are some of the topics that are discussed. This unit also contains a chapter on psychosocial aspects of sexuality.

An understanding of all of these closely interrelated phases of reproductive biology is desirable to appreciate more fully your personal life history, as well as that of generations to come.

UNIT ONE

Anatomy and Physiology of the Reproductive System

CHAPTER 2

Formation of the Egg and Sperm

Eggs and **sperm** are very special cells with a very specific purpose: to form a new individual of the next generation. The developmental history of these reproductive cells, also known as gametes, is a complex period of preparation for this special function. Discussion of the structure of gametes and the manner in which they develop is an appropriate way to begin a study of reproductive biology.

Sperm

The Testes and Their Tubules

The **testes** (also called testicles) are two ovoid bodies, about 40 mm (1½5 inches) long, externally located on the body in a sac called the **scrotum.** Each testicle has a tough external capsule of dense connective tissue. Internally, it is partitioned into several compartments. The major components of the testes are the **seminiferous tubules,** where sperm cell formation occurs. The tubules are about 1 mm in diameter (½5 inch) and are tightly coiled within the compartments (Figs. 2–1 and 2–2).

Each human testicle contains up to a half mile of seminiferous tubules. All of them terminate in a small region containing a network of passages, the **rete testis.** The sperm that are deposited in the rete by the tubules are then moved into the next portion of the male reproductive tract, the **epididymis.**

Development of Sperm

The layers of cells within a seminiferous tubule are arranged concentrically, with cells in the more mature stages located toward the center and those in

11

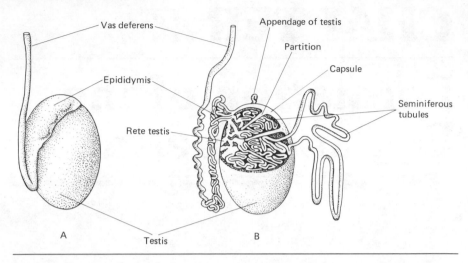

Figure 2–1 Human testes. *A,* Surface view from the side. *B,* A partial dissection. Note the seminiferous tubules tightly coiled inside several compartments. One tubule has been partially pulled out and stretched.

the more primitive stem stages around the periphery (Fig. 2–3). The stem cells are called **spermatogonia.** They are descended from cells called **primordial germ cells,** which appear very early in development, long before there is a testis (Chapters 10 and 13). These primordial germ cells migrate through the tissues of the embryo until a testicle starts to form; then they wander into the precursors of seminiferous tubules, divide many times, and remain dormant until puberty, when they start to proliferate again. Most of the cell divisions of the spermatogonia are **mitotic** (Chapter 3). Periodically, some of the stem cells start to undergo **meiosis,** the process of chromosome reduction that occurs only in gamete formation.

Those cells that have taken this first step in the formation of sperm cells now begin to move gradually toward the center **(lumen)** of the tubule. After the nuclear changes are finished, the cells continue moving toward the lumen and become transformed into mature sperm. The nucleus, the part of the cell that contains genetic material, becomes greatly condensed into a tiny package of practically pure naked deoxyribonucleic acid (DNA), or genes. A tail forms within the cytoplasm of the cell, and a special region of the head known as the **acrosome** (Chapter 10) develops. A midpiece develops between the head and tail after the tail forms, and then most of the rest of the cytoplasm is discarded. The mature sperm cell is released into the center of the tubule and is slowly transported to the rete testis and epididymis.

Structure of the Sperm Cell in Relation to Its Functions

The sperm cell is a marvel of biological engineering. A single sperm weighs only 2.86×10^{-8} mg, or 0.000000000001 oz. Despite this incredibly small

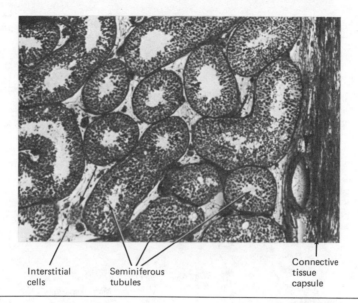

Interstitial
cells

Seminiferous
tubules

Connective
tissue
capsule

Figure 2–2 A photograph of a cross section through a testicle. Magnification
60×. (From Leeson and Leeson: *Histology,* 3rd ed. W.B. Saunders Co., Philadelphia, 1976.)

Biological material of this kind goes through an elaborate process before it
can be visualized in the manner shown. It is first placed in a solution that pre-
serves and protects all the finely detailed cellular structures. It is then embedded
in paraffin, which supports the tissue so that no shattering or distortion occurs
when it is sliced. The slicing is done on a precision machine called a microtome.
The slices are usually cut just a little more than one cell layer in thickness (10
microns). The sections are then glued to a microscope slide, the paraffin is dissolved, and the tissue is stained to bring out the features and then cleared to
make it as transparent as glass. When looking at a photograph of this nature,
remember that you are looking only at a microscopically thin slice of tissue.

The appearance of a photograph or drawing of a section will depend on (1)
the angle at which the tissue was sliced and (2) the magnification. The magnification of such an illustration is indicated by the times sign, ×: 10× means a
magnification of 10, 150× means a magnification of 150, and so on. An illustration at low magnification will show a lot of tissue and the relation of parts to
each other. An illustration at high magnification will show less tissue but will
demonstrate finer details. Compare, for example, Figures 2–2 and 2–3, which
are of the same material but at different magnifications.

This particular photograph can perhaps be visualized better if you compare
it with the appearance of a mass of hollow spaghetti that has been tightly compacted into a ball about 25 inches in diameter and frozen. A slice 0.01 inch thick
is then taken from the center and laid on a flat surface. Because the tubes of
spaghetti have been sliced at random at different angles, some will appear
round, others will appear oval, and still others will show some coiling. The seminiferous tubules of a testicle are also arranged in such a compacted mass; therefore, this low-power section shows tubules cut into different shapes.

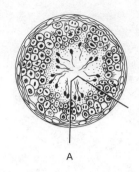

A

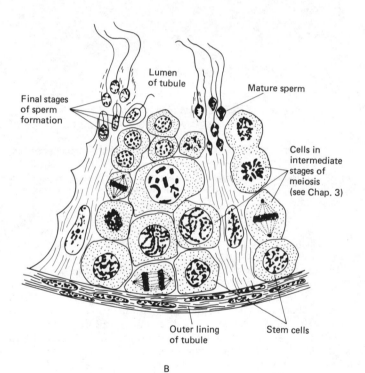

B

Figure 2–3 Sperm formation in a seminiferous tubule. *A,* Cross section of a tubule, magnification 115×. *B,* Detail at high magnification (900×) of the portion of the tubule outlined in *A.* Note (1) that the large stem cells are arranged around the periphery and (2) that as the sperm cell matures and gets smaller, it moves toward the center (lumen) of the tubule. (Adapted from Arey: *Developmental Anatomy,* W.B. Saunders Co., Philadelphia, 1965)

size, it has to be able to do a number of important things if it is to fertilize an egg successfully. First, it must get to the egg by swimming several inches through the female reproductive tract. Second, when it reaches the egg, it must interact with it in a complex fashion, since fertilization, as discussed later, is not equivalent to a simple penetration of the egg by the sperm cell. Third, and most important, this sperm cell must deliver the genetic material of the male parent into the egg. All of this is accomplished by the minute sperm cell.

Not all sperm cells look alike; there is considerable variability throughout the animal kingdom. Some are much larger than others. Some have several tails. Some swim with the aid of an undulating membrane rather than by using a tail, and the sperm cells of some crabs have a star-shaped structure and crawl rather than swim. The human sperm is actually very small by comparison to those of most other species. Even under the highest power of a light microscope, at magnifications of 500 to 1000, the head, midpiece, and tail are about all that can be distinguished (Fig. 2–4).

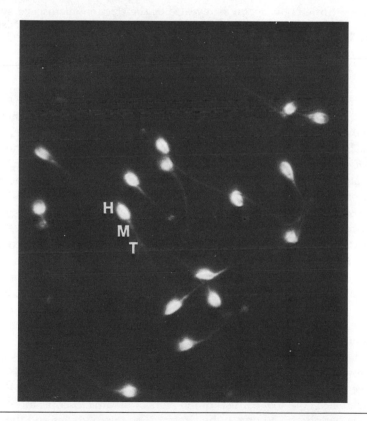

Figure 2–4 Photograph of living human sperm taken with a light (phase-contrast) microscope. Magnification 750×. The head (H), midpiece (M), and tail (T) are indicated on one of the sperm. Note that the tails and midpieces are only faintly visible, even at this high magnification.

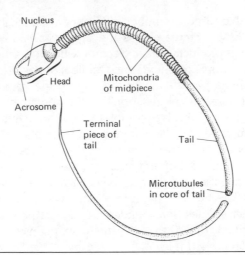

Figure 2–5 Structure of sperm as revealed by the electron microscope, magnified about 10,000×. (Adapted from Bloom and Fawcett: *A Textbook of Histology,* 10th ed. W.B. Saunders Co., Philadelphia, 1975.)

The electron microscope, which attains magnifications of 100,000 or more, reveals much more of the structure of this remarkable cell (Fig. 2–5). Under such high magnification, the head is clearly shown to be of two parts. Most of it consists of a nucleus, which contains the genetic material in a highly concentrated granular form. There is also a conspicuous cap, the acrosome, which contains substances that are released on contact with the egg cell. These substances help to digest a small region of the egg surface and thus facilitate sperm penetration.

The thickened midpiece contains a spiral-like structure that is composed of coils of **mitochondria.** Mitochondria in all cells are the organelles in which sugar is burned in a very carefully controlled manner to produce the energy that is then captured and delivered for use by other parts of the cell. Some of the mitochondria of the prospective sperm cell are packaged into the midpiece during the last stages of sperm formation to generate the energy necessary for movement once the sperm is released.

The tail of the sperm cell begins at the base of the head, extends through the midpiece, and finally terminates in an ultrathin filament. The tail contains contractile proteins similar to those found in muscle. Throughout the length of the tail there is a core that consists of ten pairs of ultramicroscopic tubules (nine around the periphery and one in the center). The microtubules and the muscle proteins are responsible for the whiplike action of the tail.

These are some of the structural features of the sperm that show how this compact, streamlined cell is able to perform its various functions.

Egg Cells

The egg, like the sperm cell, must provide one half of the genetic material for the formation of the new individual and have a structure that enables it to interact with the sperm at that very special moment of fertilization. In addition, the egg cell must supply the food material and the biochemical machinery for early development, since the sperm brings very little of these into the union. These important preparations for development are made in eggs over a period of many years.

Ovaries and Eggs

Eggs are formed in the **ovaries.** These are ovoid bodies, about 1½ inches long (40 mm), located within the female pelvis (Fig. 2–6). Each ovary consists of a central core, or **medulla,** and an outer layer, the **cortex.** The stem cells that form the eggs are found in the cortex. They are derived from the same type of primordial germ cells as the sperm cells in the male. The wandering

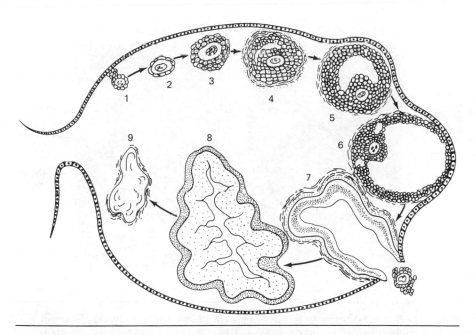

Figure 2–6 Diagram of an ovary, illustrating the development and fate of follicles. (1) Egg cell in cortex. (2 to 4) Early stages of follicle formation. (5 and 6) Maturing follicle, increasing in size mostly from fluid accumulation. (7) Ovulation. (8) Corpus luteum. (9) Degenerating corpus luteum (corpus albicans). (Adapted from Leeson and Leeson: *Histology,* 3rd ed. W.B. Saunders Co., Philadelphia, 1976.)

primordial germ cells of the very young embryo find their way into the ovary shortly after it starts to form during the second month of development. Unlike the stem cells in the testes, which continue to divide throughout the lifetime of the individual, those of the ovary proliferate very rapidly for several months and then stop dividing prior to birth. The maximum number of egg cells (**ova;** singular, **ovum**) that a woman will ever have is present three to four months before she is born. The number of eggs in the ovaries of a six-month human fetus is 7 million. This drops to 2 million at birth and 100,000 at age 20. The reasons for this great excess and steady decline in number are not known.

Follicle Formation

Eggs are very large cells. An immature egg has a diameter of 0.035 mm, five times larger than that of a red blood cell. At full maturity, the ovum is more than twice this diameter, 0.080 mm, and is one of the largest cells of the body. Ova are conspicuous in the ovary because of their size, clear cytoplasm, and extraordinarily large nuclei (see Fig. 2–8). They are found in clusters, often referred to as nests, in the cortex of the ovary (Fig. 2–7A). Starting at puberty, some ova become surrounded by groups of smaller ovarian cells called follicle

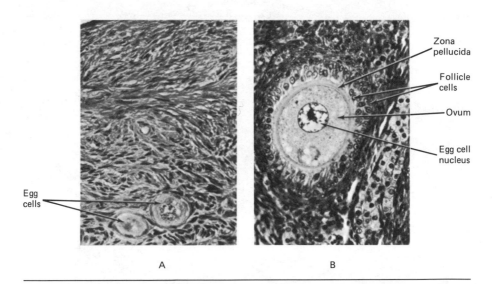

A   B

Figure 2–7 Early stages in the development of egg and follicle, as shown in photomicrographs of human ovaries. *A,* Two ova in the cortex. Magnification 125×. Note that the egg cells are much larger than any of the surrounding ovarian cells. *B,* Early stage in follicle development as the egg is now surrounded by follicle cells. Magnification 125×. The egg has also started to increase in size, as can be seen by comparing with photomicrograph *A,* which is at the same magnification. (From Leeson and Leeson: *Histology,* 3rd ed. W.B. Saunders Co., Philadelphia, 1976.)

cells. The egg, along with its surrounding cells, is now called a **follicle,** and it is within this complex that the egg finishes its maturation (Fig. 2–7B). These follicle cells contribute heavily to the final developmental stages of the egg cell. The ovum gradually grows larger as the follicle cells immediately adjacent to the egg nourish and literally inject food and other vital materials into the egg. These cells also secrete female hormones, which become concentrated in a fluid that starts to accumulate between the two layers of follicle cells (Fig. 2–8A). As the follicle enters its last stage of maturation, usually once a month during the woman's reproductive lifetime, this fluid accumulates rapidly, expanding the follicle and pushing the egg to one side. The egg at this stage is embedded in a hill-like column of follicle cells. One region of this follicle reaches the surface of the ovary and becomes very thin as the midpoint between the two menstrual periods approaches. The pressure of the accumulating fluid increases, and the follicle ruptures, expelling the egg along with the hormone-rich fluid (Fig. 2–8B). The released ovum is still surrounded by a

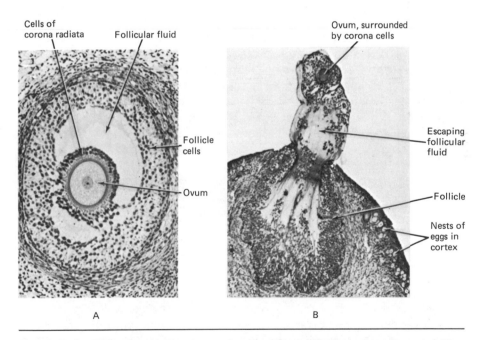

Figure 2–8 Follicular maturation and ovulation. *A,* Nearly mature human follicle. Magnification 60×. (From Leeson and Leeson: *Histology,* 3rd ed. W.B. Saunders Co., Philadelphia, 1976.) *B,* A remarkable photograph of a rabbit ovary caught at the precise second at which ovulation was occurring. Magnified about 20×. Note the stream of follicular fluid that is escaping along with the ovum and its surrounding cells. (From Page, Villee, and Villee: *Human Reproduction,* 2nd ed. W.B. Saunders Co., Philadelphia, 1976. Courtesy of R.J. Blandau.)

layer of cells, which is now known as the **corona radiata** because of its crownlike appearance (see Fig. 10–1).

The job of the versatile follicle cells that both nourished the egg and secreted some hormones is still not completed. The cells that remain in the deflated follicle undergo a change and form a bright yellow spherical structure called the **corpus luteum** (see Fig. 2–6). The corpus luteum exists for at least two more weeks, and the cells within it now secrete a combination of sex hormones different from those they secreted earlier in the cycle. If pregnancy occurs, the corpus luteum continues to grow to a rather large size and manufactures hormones that are very important for maintaining pregnancy. The corpus luteum simply degenerates (Fig. 2–6) if no pregnancy occurs. The nature of the hormones formed in the ovary and their various functions are discussed in Chapter 8, but the cyclical changes in the structures that secrete them should already be apparent. Specifically, these changes are (1) the preovulatory accumulation of follicular fluid and its sudden release at ovulation and (2) the postovulatory growth and decline of the corpus luteum.

Structure of the Egg

The newly released ovum is an exceptionally large cell with a huge nucleus. This nucleus is abnormally large because it contains a high concentration of **nucleic acids.** Their synthesis and storage in the nucleus are vital aspects of the prolonged development of eggs, just as important as the synthesis and storage of food material that occurred in the cytoplasm. Some of these nucleic acids are molecules of **messenger RNA** (ribonucleic acid), which carry coded information, or instructions, from the genes of the nucleus to the cytoplasm. Other nucleic acids (e.g., those in ribosomes) are used to set up rapidly a protein-synthesizing system in the cytoplasm after fertilization. These stored nucleic acids are released into the cytoplasm after the egg is discharged and the cell goes into the final stages of nuclear maturation.

Remember that the ovum is a unique cell, the only cell of the body that has the capacity of forming a new individual of the next generation. This can occur only if many things have gone right during the 20 to 30 years in which the cell was developing. This preparatory period and its significance is discussed in more detail in the chapters on development.

There is one more idea to ponder, especially in this era of concern about our environment and the effects of its pollution. The egg cell is present for many years before it is released. The follicular environment is the first one to which a new individual is exposed, long before fertilization, and it is not immune to pollution. The factors that can affect the quality of life start to operate in the ovary. This feature of development is also examined in more detail further on.

CHAPTER 3

Chromosomes and Sex

Chromosomes

The adult human body is composed of trillions of **cells,** the basic biological units of life. All cells have two main components: **cytoplasm** and a **nucleus** (Fig. 3–1). The cytoplasm is the peripheral portion where the actual work of the cell is performed, where the sugar we ingest is transformed into energy and where proteins such as hemoglobin and digestive enzymes are synthesized. The nucleus is usually located in the center. It contains the hereditary material of the cell and controls the metabolic activities occurring in the cytoplasm.

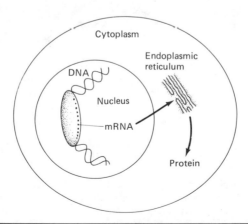

Figure 3–1 Diagram of a cell showing how the nucleus controls protein synthesis in the cytoplasm.

The units of heredity are called **genes.** These are specially coded segments of a large-chain molecule called deoxyribonucleic acid, or DNA, which is normally coiled as a double helix. Only a small portion of the numerous genes in every nucleus are functioning, that is, working, at any one time. Various events within the cell or outside it determine whether a gene is switched on or off at any given time. When a gene is switched on, the segment of DNA where it is located unwinds, and the coded message that this region contains is transcribed onto another molecule, called messenger ribonucleic acid (mRNA). The mRNA then moves out of the nucleus and into the cytoplasm, where it becomes associated with an organelle (called the endoplasmic reticulum) whose function is to form **protein** molecules, the molecules that perform most of the work of a cell. It is here that the message from the gene becomes translated into the synthesis of a specific protein (Fig. 3–1).

Of all of the approximately 160 different types of cells that compose the

Figure 3–2 A set of chromosomes from a human cell. The most propitious time to examine them is during cell division, after the chromosomes have duplicated but before they have separated. They appear in various X-shaped configurations because the members of a duplicated chromosome are still attached to each other at one point. (From Mange and Mange: *Genetics: Human Aspects.* Saunders College Publishing, Philadelphia, 1980.)

human body, the egg cell is the only one that has the capacity to form a new individual. Although the egg needs that critical meeting with a sperm in order to do so, the egg cell is the primary intermediary between two generations of adults. Since we have dismissed the convenient but false notion that there is a tiny preformed individual in the egg, ready simply to expand, we must then look upon the egg cell as a complex package of coded information that becomes translated into a new person during the developmental period. Some of this information is in the cytoplasm of the egg, deposited there during its long stay in the ovary. But when coded information in a cell is mentioned, we usually think of the nucleus and the genes that it contains.

Genes are contained within the **chromosomes** found in the nucleus. This word comes from Latin *chromo,* meaning color, and *soma,* meaning body. They were called "colored bodies" by the early microscopists because they stain brilliantly with the dyes that biologists use to make cell structures visible (Fig. 3–2). All of the cells of an individual have the same number and kinds of chromosomes, a fact that is consistent with their great importance as carriers of hereditary material. Each cell of a dog contains 78 chromosomes. Each cell of a cat contains 38. Every cell of a human being contains 46 chromosomes. The chromosomes in a cell are matched as pairs (Fig. 3–3). At the

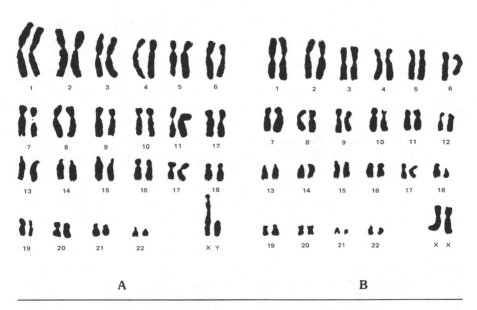

A   B

Figure 3–3 Human chromosomes of *A,* a normal man, and *B,* a normal woman, arranged in the conventional manner known as a karyotype. The chromosomes in a photograph of a preparation, such as that of Figure 3–2, are cut out with scissors, and the 44 autosomes are then arranged in pairs according to decreasing length and morphological similarities. The sex chromosomes are located on the bottom right. (From Page, Villee, and Villee: *Human Reproduction,* 2nd ed. W.B. Saunders Co., Philadelphia, 1976. Courtesy of Dr. M. Grumbach.)

time of fertilization, one member of a pair comes from the maternal parent, and the other comes from the paternal parent. The integrity of this chromosome complex is maintained throughout all of the millions of cell divisions that occur in the formation of an adult from a single cell, the fertilized egg.

Chromosomes in Ordinary Cell Division

Every time a cell divides, a series of elaborate mechanisms operates (1) to ensure a precise molecule-by-molecule duplication of each chromosome and (2) to distribute the duplicated chromosomes in an orderly fashion to each of the daughter cells so that the constancy of the complex is maintained. This process is called **mitosis** (Fig. 3–4). In its simplest essence, mitosis consists of a single cytoplasmic division and a single duplication of chromosomes. There is a more or less equal division of cytoplasm between the two new cells. The duplicated chromosomes are equally distributed between the daughter cells, and the chromosomal complex of each is precisely the same, right down to every single gene.

Chromosomes in the Formation of Egg and Sperm

In the development of gametes, it is obvious that something different must occur as far as the chromosomes are concerned. If human sperm and egg were both to contain 46 chromosomes and these two cells would then fuse, cells of the next generation would contain 92 chromosomes, those of the generation following that would have 184 chromosomes, and so forth—an impossible situation. During the formation of the gametes, the chromosomal number is halved, so that the number of chromosomes characteristic of the species is restored when sperm and egg fuse at fertilization.

This special type of division of cells that form the gametes is called **meiosis.** During meiosis, there are two cytoplasmic divisions but still only one replication of the chromosomes. The net result is that four cells are formed, each with half the normal chromosomal number (Fig. 3–5). This is not a random half. The chromosomal divisions and distributions are such that only one member of each pair ends up in each of the four cells that will form a gamete. Thus when a human sperm and egg unite at fertilization, it is not just 46 chromosomes that are present in the new individual, but the 23 pairs that are characteristic of the species.

In the formation of ova, there is a minor variation in the typical pattern of meiosis in that only one of the four cells that form goes on to develop into an egg. A relatively large amount of food and other material important for development accumulates in a structural pattern in the cytoplasm of the egg during its formation. Meiosis occurs late in the history of the egg, and most of

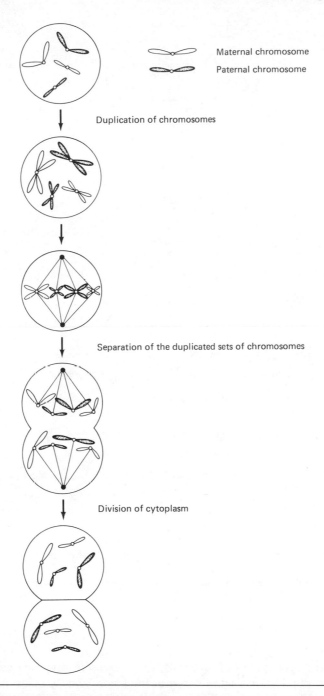

Maternal chromosome
Paternal chromosome

Duplication of chromosomes

Separation of the duplicated sets of chromosomes

Division of cytoplasm

Figure 3–4 Diagram of mitosis in a cell containing two pairs of chromosomes. There is a single duplication of chromosomes and a single cell division. The two daughter cells are fully equivalent genetically to the cell from which they arose.

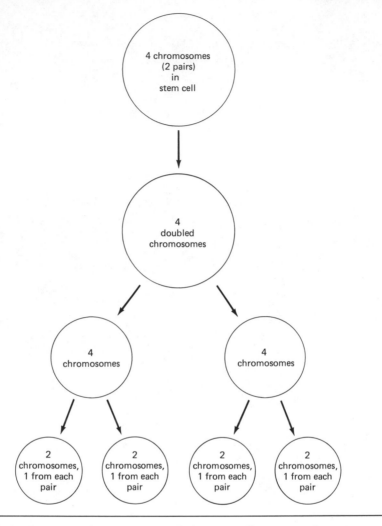

Figure 3–5 Diagram of meiosis in a cell that normally contains four chromosomes, that is, two pairs. There is a single duplication of chromosomes in the stem cell followed by two cell divisions. The net result is the formation of four cells, all containing two chromosomes, one member of each pair.

this important aggregation of materials is conserved in only one cell. There is still but one duplication of the chromosomes and two cytoplasmic divisions, but those divisions are unequal. The end result is one large egg cell and three small, insignificant **polar bodies** (Fig. 3–6).

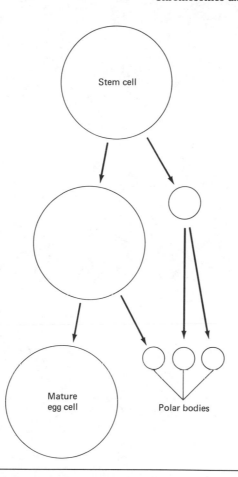

Figure 3–6 Diagram of meiosis in egg cells. Almost all of the cytoplasm is conserved in one of the four cells, and that one is the egg. All three polar bodies, however, have nuclei that are equivalent to that of the egg.

Chromosomes and Sex Determination

Of the 46 chromosomes in each human cell, 22 are pairs of regular chromosomes known as **autosomes,** and the other 2 are the **sex chromosomes** (Fig. 3–3). These latter two differ in the two sexes. In every cell of a woman, there are 22 pairs of autosomes plus a pair of sex chromosomes known as **X chromosomes.** Every cell of a man, on the other hand, contains 22 pairs of autosomes and an unmatched pair of sex chromosomes, an X and a **Y chromosome.** When chromosomal reduction occurs in meiosis, each egg cell will contain 22 autosomes (single, not paired) and a single X chromosome (Fig. 3–7). All egg cells will be alike in this respect. However, when meiosis occurs in

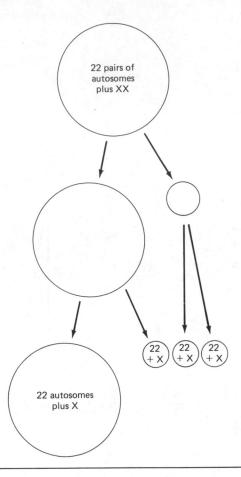

Figure 3–7 **The sex chromosomes in the formation of human eggs. No matter how reduction occurs, all eggs end up containing an X chromosome.**

sperm cells, the unmatched sex chromosomes will each end up in a different sperm cell. Therefore, half of these sperm cells will contain an X chromosome, and the other half will contain a Y chromosome (Fig. 3–8). In terms of chromosomal composition, all egg cells are alike, but sperm cells are of two types.

At the moment of fertilization, the eventual chromosomal composition of the cells of the embryo will depend on which type of sperm fertilizes the egg. If an X-bearing sperm cell fertilizes the X-bearing egg, the new cell will contain 22 pairs of autosomes plus two X chromosomes, one from the egg cell and one from the sperm (Fig. 3–9). This fertilized egg will develop into a female. If, on the other hand, a Y-bearing sperm cell fertilizes the X-bearing egg, the new cell will also contain 22 pairs of autosomes, but the sex chromosomes will consist of an X and a Y combination. This fertilized egg develops into a male.

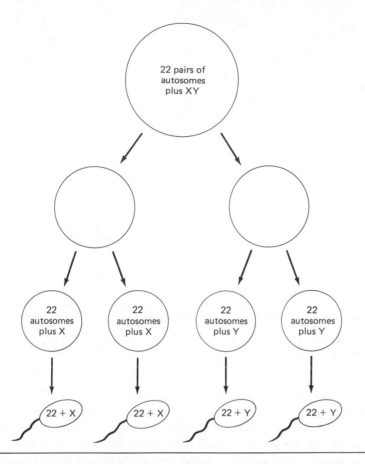

Figure 3–8 The sex chromosomes in the formation of human sperm. During the reduction of chromosome number, half of the sperm will carry an X chromosome, and half will carry a Y chromosome.

In this manner, the sex of an individual is established at the moment of conception, depending on whether an X-bearing or a Y-bearing sperm fertilizes the egg.

This process of genetic sex determination has several important implications. It is possible that a practical method for separation of X- and Y-bearing sperm may be developed in the near future, and therefore control of the sex of offspring will be feasible (Chapter 20). It is now practical to determine the genetic sex of an individual, even in cases in which the results of a physical examination may be ambiguous. This can be done (1) by examining the chromosomes, a precise but somewhat difficult technique (Fig. 3–3), or (2) by examining nuclei in a sample of white blood cells, which, if they contain two X chromosomes, often display visible evidence of them. The latter technique,

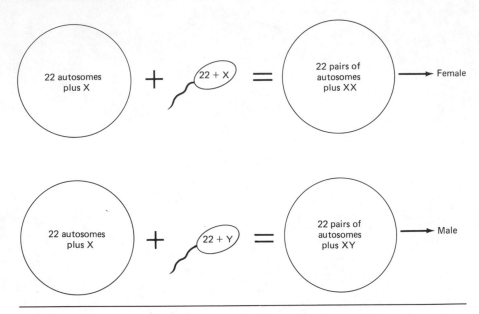

Figure 3–9 The chromosomal basis of sex determination at the time of fertilization. Since all eggs carry an X chromosome, the combination in the fertilized egg will depend on whether an X-bearing or a Y-bearing sperm fertilizes the egg.

in particular, is relatively simple and is becoming frequently used in international athletic competition to ensure that all athletes competing as women are truly genetic females and not genetic males who have a feminine appearance.

In summary, remember that (1) the genes and the chromosomes that contain them are very important for every cell and that a precise duplication, molecule for molecule, occurs at each cell division; (2) during maturation of the sex cells, this chromosome number is reduced to one half; (3) chromosomes are responsible for the sex of the individual; and (4) this sex is established at the moment of conception.

CHAPTER 4

Anatomy and Physiology of the Male Reproductive Organs

Histology: An Introduction to the Microscopic Architecture of Organs

An **organ** is an anatomical unit of the body having a recognizable form and one or more specific functions. The testis, uterus, bladder, and clitoris are examples. Organs are always composed of several kinds of **tissues.** A tissue is a group of cells similar in structure and function. **Muscle, nerve, epithelia,** and **connective tissue** are the four basic tissue types in the human body, and all four are found in virtually all organs. **Histology** is the study of the microscopic structure of tissues, of the manner in which they are arranged into organs, and of how tissue structure is related to the function of organs. It is, therefore, desirable to review the types of tissue and see how they interrelate in a typical organ before studying the reproductive organs.

The Basic Tissues

Epithelia   These tissues consist of a group of cells usually arranged in a continuous layer for the purpose of lining or covering something. The cells of an epithelial layer may be flat (Fig. 4–1A) or columnar (Fig. 4–1C) and may be a single layer in thickness or stratified into many layers (Fig. 4–1B). Some epithelial cells contain numerous **cilia** on their free surface. These are microscopic eyelash-like projections that synchronously beat in waves and have the capacity to propel small objects like egg cells along the oviducts (Fig. 4–1D). A layer of epithelium lines all surfaces and cavities, large and small, internal and external. The skin and the linings of the mouth and intestine are examples.

The major function of epithelial tissues is to protect the surface that they

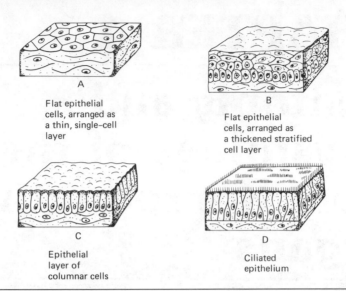

A

Flat epithelial
cells, arranged as
a thin, single–cell
layer

B

Flat epithelial
cells, arranged as
a thickened stratified
cell layer

C

Epithelial
layer of
columnar cells

D

Ciliated
epithelium

Figure 4–1 Epithelial tissues. (Adapted from Jacob, Francone, and Lossow: *Structure and Function in Man,* 4th ed. W.B. Saunders Co., Philadelphia, 1978.)

cover. They also have one other capacity, namely, to secrete substances. A group of epithelial cells that secrete is called a **gland.** A typical gland develops from a layer of cells by forming an inpocketing (invagination), growing internally, and branching (Fig. 4–2). The cells of the terminal ends of the pocket eventually secrete the fluid; other cells line the ducts by which the secretion is delivered to the epithelial surface. Glands may be small, such as the numerous sweat glands of the skin; or they may be quite large, such as the salivary glands, the prostate gland, which secretes the nutritive fluid for sperm, and the mammary glands, which secrete milk. Large or small, all glands start as simple inpocketings of a group of epithelial cells.

Connective Tissue This tissue provides support to organs and other tissues by means of interlacing fibers. Connective tissue also contains cells that secrete the fibers and have a few other functions. **Areolar connective tissue** fills spaces and provides for loose attachments of tissues (Fig. 4–3). It is composed of a light network of fibers interspersed with cells that contain a large amount of fat. Connective tissue fibers that are arranged into sheets cover bundles of nerves, muscles, and other organs (Fig. 4–5B). Extra-strong layers of this type are called fascia. Very dense collections of connective tissue fibers form the dermis of the skin (that part that can be tanned into leather), tendons (which connect muscle to bone), and ligaments (which connect bone to bone). Bone is a dense arrangement of connective tissue that has become infiltrated with minerals.

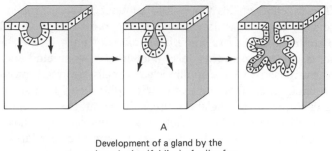

A

Development of a gland by the
inpocketing (folding) of cells of an
epithelial tissue.

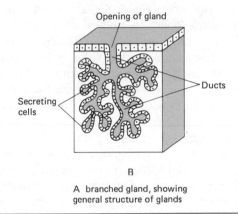

B

A branched gland, showing
general structure of glands

Figure 4–2 Glands: their development and structure. (Adapted from Jacob, Francone, and Lossow: *Structure and Function in Man,* 4th ed. W.B. Saunders Co., Philadelphia, 1978.)

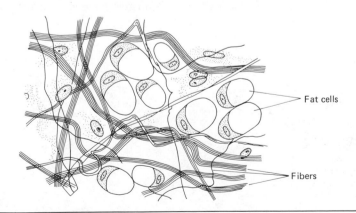

Figure 4–3 Areolar connective tissue.

Muscle The function of this tissue is to contract. Special protein molecules within these cells, arranged as longitudinal fibers, provide muscle cells with the capacity to contract. There are three basic types of muscle: skeletal, cardiac, and smooth. **Skeletal muscles** (also known as cross-striated and voluntary muscle) are composed of large multinucleated cells arranged into long fibers (Fig. 4–4A). Their function is to move a part of the skeleton or some soft part, such as the vaginal orifice. Skeletal muscle can contract very rapidly. It is ordinarily under voluntary control, although some involuntary reflexes are mediated by skeletal muscle (e.g., breathing and some phases of orgasm). **Cardiac muscle** is similar in structure to skeletal muscle but is found only in the heart. **Smooth muscle** (also known as visceral and involuntary muscle) is composed of spindle-shaped cells (Fig. 4–4B) that are found in small clusters or as layers. Smooth muscle fibers contract much more slowly than skeletal muscle fibers and often in a rhythmic fashion. Their contraction is not ordinarily under conscious voluntary control. They form an integral part of the intestine, arteries, uterus, oviducts, and other hollow internal organs. Visceral muscle cells are usually arranged in two or more layers lying in different planes, which permits a hollow organ to contract in more than one plane (Fig. 4–6).

Nerve Neurons (nerve cells) are specialized for one function: to conduct impulses. The nucleus and most of the cytoplasm are located in one region, called the cell body (Fig. 4–5A). Extending from this cell body is a long appendage, the axon, which may be several feet long. The terminal end has a plate by which the axon is attached to a muscle or sense organ. The cell body also has numerous other extensions, called dendrites, by which it can receive stimuli and interconnect with other nerve cells. Cell bodies of neurons are almost always clustered within or adjacent to the brain and spinal cord. Their axons can extend within the central nervous system or out to all parts of the body. Peripheral nerves are bundles of axons held together by connective tissue (Fig. 4–4B). The function of nerve cells is to transmit messages. Sensory nerves

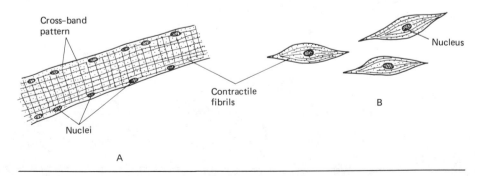

Cross–band pattern

Nucleus

Contractile fibrils

B

Nuclei

A

Figure 4–4 Muscle tissue. *A,* A single fiber of skeletal muscle. *B,* Several visceral muscle cells.

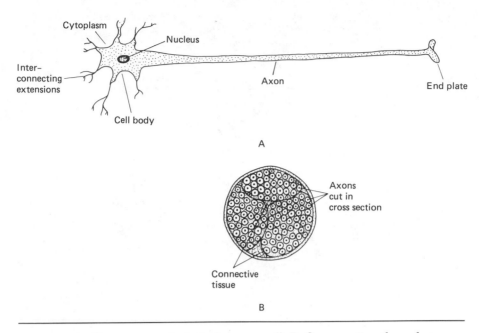

Figure 4–5   Nerve tissue. *A,* A single nerve cell. *B,* Cross section through a peripheral nerve bundle.

carry messages of pain, touch, and temperature to the brain and spinal cord. Interneurons transmit messages between various portions of the brain and spinal cord. Motor nerves carry messages from these areas and stimulate muscles to contract and glands to secrete.

Arrangement of Tissues in a Typical Internal Organ

The basic tissues can be combined in various ways to form different kinds of organs. Many of the reproductive organs to be studied have a similar architecture. Viewed in cross section, that is, perpendicular to the long axis, a hollow organ can be described as follows (Fig. 4–6): The center, or **lumen,** of any hollow organ is lined with an epithelium, which is almost always associated with some kind of gland. The glands have grown from the epithelium into a matrix of areolar connective tissue. They secrete their product into the lumen via ducts. Two layers of smooth muscle usually surround a hollow organ. The circular layer, composed of cells whose long axes are parallel to the circumference of the organ, will constrict the diameter of the tube when it contracts. The longitudinal layer, composed of cells whose long axes are parallel to the long axis of the organ, will shorten the length of the tube when it contracts. There is usually a thin layer of areolar connective tissue external to these two layers. The outside of the organ is covered by another layer of epithelium. Blood vessels and nerves are interspersed throughout.

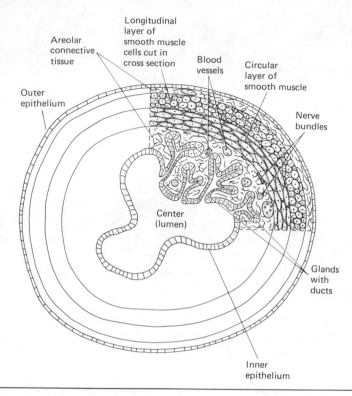

Figure 4–6 Arrangement of tissues in a typical internal organ.

This description is a simple example of microscopic architecture, of the manner in which cells and tissues are arranged to form an organ.

Anatomy of the Pelvis

The **pelvis** is a region of the body that contains many reproductive organs. It is a funnel-shaped structure composed of four bones: (1) **sacrum,** by which it is attached to the rest of the backbone; (2) **ilium;** (3) **ischium;** and (4) **pubis** (Figs. 4–7 and 4–8). The last three are fused together to complete a cone. The **femur** (thigh bone) fits into a socket formed by these three. Some landmarks that can be easily **palpated** (felt with the fingers) from the outside are (1) the **iliac crest,** the flared brim of the pelvis that marks the outside of the hip; (2) the **ischial tuberosities,** which are the bones upon which you sit; and (3) the **pubic arch,** which runs across the front of the abdomen. Locate these on yourself. The **coccyx** and **ischial spines** are internal landmarks of importance because they project into the pelvic outlet and thus limit the size of the birth canal in women. The two pubic bones are joined together in front by a tight ligament, the **pubic symphysis.**

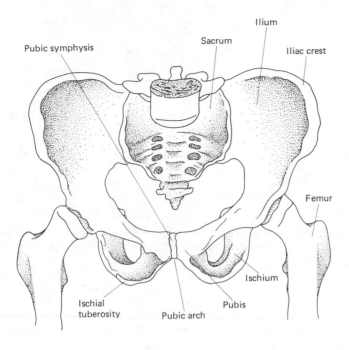

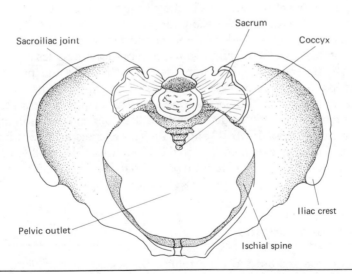

Figure 4—7 **The human pelvis, as viewed from the front (A) and from above (B).**

Most of the bones of the body show some sexual differences. The pelvis, however, is conspicuously different in men and women (Fig. 4—8). In women (1) the pubic arch is broader; (2) the iliac crest is flared to a greater degree, making the hips broader; and (3) the coccyx and ischial spines do not impinge as much into the **pelvic outlet** as in men. The net result of these and other

Female   Male

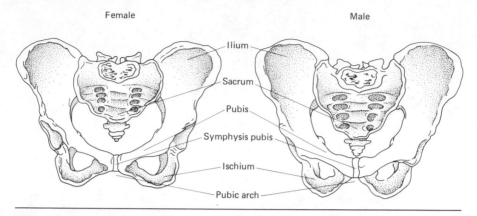

Figure 4–8   A comparison of typical female and male pelvic bones. (Adapted from Dienhart: *Basic Human Anatomy and Physiology.* W.B. Saunders Co., Philadelphia, 1967.)

differences is that this outlet, which is important during delivery, is significantly larger in women. Its boundaries are carefully measured by the obstetrician during early pregnancy.

The pelvic outlet is a structurally strategic area. It must support the weight of all the abdominal organs, yet allow for passage of babies and of feces and urine. The outlet can be divided into front and rear portions. The larger posterior portion contains the rectum and a group of strong muscles and ligaments (see Figs. 4–10 and 5–2). The front portion, which fills the pubic arch, is supported by the **urogenital diaphragm,** which contains the urethral opening and, in women, the vaginal opening. This diaphragm consists of two layers of strong connective tissue attached to the front and rear portions of the pubic arch (see Figs. 4–10, 5–1, and 5–2). These layers are spaced about ½ inch apart. The sphincter muscles for the urethra and the vagina as well as several other small structures are sandwiched between these layers of connective tissue.

The Male Sex Organs

The male sex organs are the (1) scrotum, (2) testes, (3) epididymis, (4) vas deferens, (5) spermatic cord, (6) seminal vesicles, (7) prostate gland, (8) Cowper's glands, (9) urethra, and (10) penis.

The principal functions of the testes are to form sperm cells and to secrete the male hormones. Sperm formation occurs in the many yards of seminiferous tubules tightly coiled within the testicular compartments. The internal structure of these has been discussed in Chapter 2. The other major function of the testes, hormone secretion, is performed by rather modest-looking aggregations of cells in the interstices between the seminiferous tubules. They are called **interstitial cells** (Figs. 2–2 and 4–9).

The testes are located externally in the **scrotum,** a pouch of skin, muscle, and fascia (sheets of connective tissue). This pouch also serves to regulate the temperature of the testes (p. 46). Scrotum and testes are well endowed with pain nerves, one of nature's ways of protecting important exposed organs. The testes are enclosed by a tough connective tissue capsule called the **tunica albuginea** ("white cloak"). This strong enclosure does not permit the testes to expand much, which is why any injury to the testes that causes inflammation and swelling is very painful.

All of the coils of seminiferous tubules terminate in one region of the testes in an irregular network of tubules called the **rete testis** (Figs. 2–1 and 4–10). Mature sperm cells collect here before being passed on to the next part of the male reproductive tract, the epididymis.

The **epididymis** is a cap-shaped structure along the side of the testis consisting of some coiled tubules and a duct (Figs. 2–1 and 4–10). This is the region where sperm are stored and continue a physiological maturation, or ripening, that is necessary before they can fertilize an egg.

The epididymis leads into the principal reproductive duct of the male, the **vas deferens.** This tube is about 15 inches (48 cm) long and ¼ inch (7 mm) wide. It is a solid muscular tube with only a small internal space (Fig. 4–11).

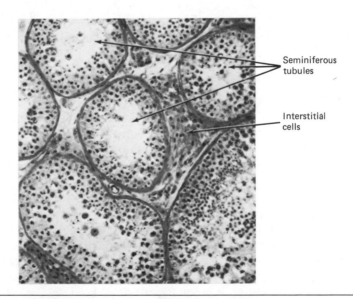

Seminiferous tubules

Interstitial cells

Figure 4–9 **Photograph of a cross section of a human testicle showing interstitial cells. Magnification 120×. (From Bloom and Fawcett:** *A Textbook of Histology,* **10th ed. W.B. Saunders Co., Philadelphia, 1975.)**

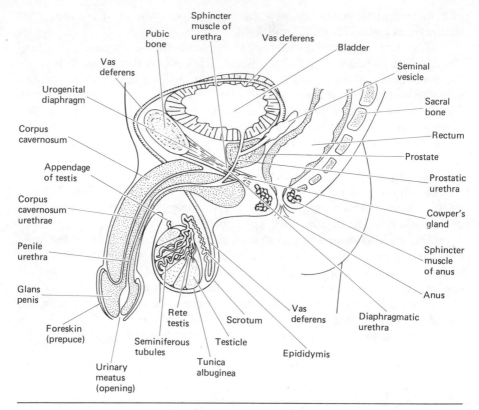

Figure 4–10 Diagram of the male reproductive organs in sagittal section.

The microscopic structure resembles that of the "typical" organ described earlier in this chapter (see Fig. 4–6). The vas deferens extends from the scrotum and over the front rim of the pubic bone as the main component of the **spermatic cord** (Fig. 4–16). It then penetrates the abdominal wall through the inguinal canal and enters the pelvis to join the urinary system just below the bladder (Fig. 4–10).

The **ampulla** is a swelling of the vas deferens just before it joins the urethra. The **seminal vesicles** are outgrowths from the base of the vas deferens (Fig. 4–10). They do not store sperm, as their name implies, but are a pair of glands that secrete 30 percent of the fluid portion of semen.

The **prostate** is a gland that secretes 60 percent of the fluid of semen. Often described as having the size and shape of a chestnut, it surrounds the urethra just below the bladder. It is not a single gland but rather a collection of about 60 glands, each with a separate opening into the urethra (Figs. 4–10 and 4–12).

Cowper's glands are a pair of small glands located within the urogen-

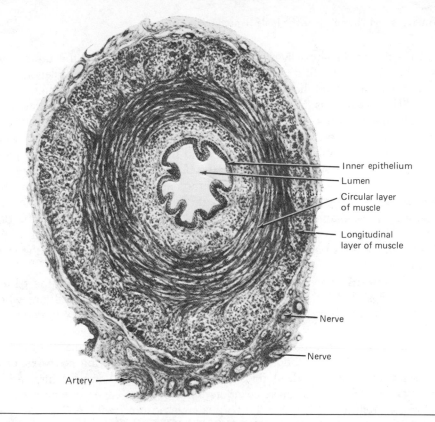

Inner epithelium

Lumen

Circular layer
of muscle

Longitudinal
layer of muscle

Nerve

Nerve

Artery

Figure 4–11 Photomicrograph of a human vas deferens in cross section. Magnification 100×. (Adapted from Bloom and Fawcett: *A Textbook of Histology*, 10th ed. W.B. Saunders Co., Philadelphia, 1975.)

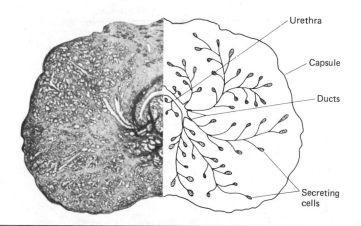

Urethra

Capsule

Ducts

Secreting
cells

Figure 4–12 The prostate gland. The left side is a photograph, and the right is the diagram. Note how the urethra is surrounded by glandular tissue. (Photograph from Bloom and Fawcett: *Textbook of Histology*, 10th ed. W.B. Saunders Co., Philadelphia, 1975.)

ital diaphragm. They deliver a viscous, sticky secretion into the urethra via small ducts (Fig. 4–10) during sexual excitement and prior to ejaculation, neutralizing the acidity of the urethra and preparing the way for the semen.

The **urethra** is the tube that extends from the bladder and carries urine for elimination. It also serves in men as the final passageway for the products of the reproductive system. It is approximately 4 inches (10 cm) long and has three portions: prostatic, diaphragmatic, and penile (see Fig. 4–10). The prostatic portion is the segment just below the bladder that is surrounded by prostatic tissue (Fig. 4–12). The short diaphragmatic portion is the segment that leaves the pelvis by passing through the urogenital diaphragm. This is the only opening in the diaphragm in men. The urethra is surrounded within the diaphragm by some small muscles that act as a sphincter to open and shut off the flow of urine. The penile portion, which is the longest, goes through the shaft of the penis.

The **penis** consists of a shaft and an enlarged head called the **glans penis** (Figs. 4–10 and 4–13). It extends 2 to 4 inches (5 to 10 cm) when flaccid and can become 5 to 7 inches (13 to 18 cm) long when engorged and erect.

The penis is composed of three cylinders of spongy vascular tissue, each covered with a layer of tough connective tissue. These are the **corpora cavernosa** (cavernous bodies) (Figs. 4–13 and 4–14). A large central artery courses through each of these cylinders. One of these cavernous bodies contains the urethra and is called the corpus cavernosum urethrae. This structure originates internally in an enlarged bulb that is attached to the urogenital dia-

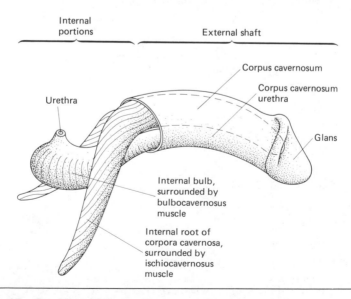

Figure 4–13 Structure of the penis.

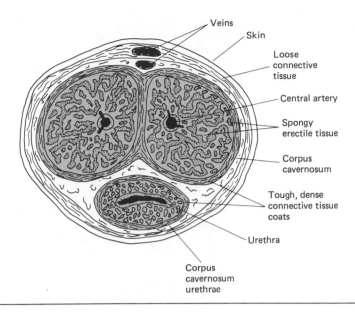

Veins

Skin

Loose
connective
tissue

Central artery

Spongy
erectile tissue

Corpus
cavernosum

Tough, dense
connective tissue
coats

Urethra

Corpus
cavernosum
urethrae

Figure 4–14 Structure of the penis in cross section.

phragm. At this internal base, it receives the urethra. Then both the cavernous body and the urethra extend the length of the penis, forming the lowermost compartment of the shaft. The corpus cavernosum urethrae is expanded at the tip, forming the glans penis. The two upper corpora cavernosa originate just behind the glans and extend across the pendulous portion of the penis. Internally, they separate, move laterally and deep (Fig. 4–13), and become firmly anchored to the pelvic bone (the ischial portion of the pubic arch). The internal portions of the penis are surrounded by thin sheets of muscle. The bulb of the corpus cavernosum urethrae is covered by the **bulbocavernosus muscle,** and the lateral roots are covered by the **ischiocavernosus muscle.** The volume of these internal portions is considerable, about equal to that of the external pendulous portion.

The expanded terminal portion of the penis, the glans, is the most sensitive portion of the shaft. The rim is particularly well innervated by sensory neurons. The skin that covers the glans is loose and not attached, except at the base behind the rim (Figs. 4–10 and 4–15). This is the **foreskin,** or prepuce. The foreskin of Jewish boys is ritually cut off by a procedure called circumcision. To facilitate sanitation, circumcision has also become very popular among non-Jewish groups in the United States but not in other countries (e.g., only 6 percent of Canadian boys are circumcised). This practice in the United States is now being questioned, since (1) the American Academy of Pediatrics has labeled circumcision as unnecessary surgery; (2) this surgery is rarely performed with the benefit of anesthetic and sometimes causes compli-

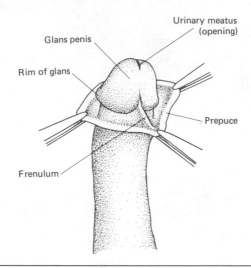

Figure 4–15 Base of the penis with the foreskin retracted.

cations afterward; and (3) lack of circumcision does not create a sanitation problem, since the foreskin can be readily rolled back in young boys and men and the glans area can be washed.

The penis becomes erect during sexual excitement because specific nerves cause the large arteries in the centers of the corpora cavernosa to dilate and deliver rapidly a fair amount of blood into the spongy tissue of the corpora. Simultaneously, the veins that drain blood from the penis are prevented from doing so by a reflex closure of valves within them. The corpora, as a consequence, elongate and expand rapidly to the limits imposed on them by their tough coats of connective tissue and become turgid. The turgidity of the well-anchored internal portions of the corpora causes the entire penis to become erect. This hydraulic effect can be compared to some degree with the increase in turgidity that occurs in a garden hose when the water pressure is turned on but escape is prevented by closing the nozzle. Contrary to common belief, there is no bone in the human penis to help maintain an erection, although a penile bone, the baculum, is found in some animals such as dogs, bats, and monkeys. After ejaculation (which is discussed in Chapter 6), the veins in the penis open up and start to drain the excess volume of blood. This quickly reduces the size and turgidity of the penis.

Composition of Semen

Semen is the material ejaculated during the male orgasm. The volume ranges from 2 to 6 ml (about ⅕ oz). Sperm cells are its most important constituent but compose only 10 percent of the total volume. Semen usually contains between 80 to 100 million sperm per milliliter, but a concentration range of 40 to 200 million per milliliter is still considered normal. This means that a single ejaculate may contain anywhere from 80 to 1200 million sperm cells. Why is there such a tremendous surplus when only one is needed to fertilize? This is not fully understood. Perhaps many are needed because the hazards of the trip through the female reproductive tract are great and few sperm actually complete it, or perhaps many sperm are necessary to help dissolve the coatings around the egg. It is known, however, that concentrations of less than 40 million sperm cells per milliliter in the ejaculate are associated with male infertility (see Chapter 18).

The fluid secreted by the seminal vesicles constitutes about 30 percent of the ejaculate. The main ingredient is sugar (fructose) for energy. The other important constituents are several kinds of **prostaglandins,** important substances that stimulate contraction of the smooth muscle of the uterus and oviducts and thus help to propel sperm toward the ovary (see Chapters 5, 15, and 18).

Prostatic fluid constitutes 60 percent of the ejaculate. It is very viscous and is composed of several enzymes (acid phosphatase, muramidase, and others), citric acid, and zinc. The precise functions of all of these components are not known. They are, however, quite necessary; it is known that prostatic malfunction can cause male infertility. The secretions of all the male accessory glands facilitate sperm action by (1) stimulating sperm motility, (2) providing nutrients, and (3) helping to neutralize the chemically hostile environments of the male urethra and the vagina.

The Spermatic Cord and the Inguinal Canal

The testes develop high within the abdomen, next to the kidneys, through most of the fetal period. They migrate into the scrotum during the seventh to ninth months of pregnancy. They slowly descend toward the pelvis, travel through an oblique passageway in the abdominal wall, the **inguinal canal,** and pass over the rim of the pubic bone and down into the waiting scrotum. The testes tow the vas deferens, nerves, arteries, and veins with them during this descent. As they migrate through the various layers of muscle and fascia of the abdominal wall, the testes also pick up portions of these layers and carry them into the scrotum. Consequently, both testes and vas deferens are surrounded by layers of muscle and fascia that are continuous with those of the abdominal wall.

The **spermatic cord** is a complex of structures that extends between the testes and the inguinal canal. The spermatic cord is about ½ inch (12 mm) thick and contains the vas deferens, nerves, an artery, and a venous network called the pampiniform plexus (Fig. 4–16). It can be palpated by gently moving the fingers over the pubic bone close to the midline. The spermatic cord is

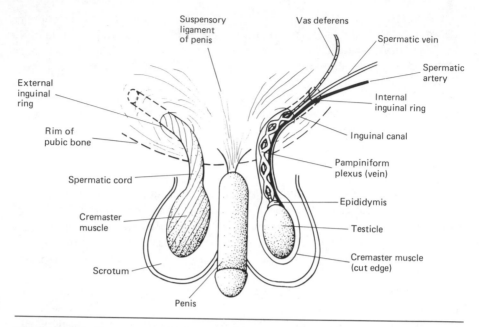

Figure 4–16 Structure and relationships of the spermatic cord. A superficial dissection is shown on the left, and the internal contents are shown on the right.

covered by a layer of fascia and muscle known as the **dartos,** which helps to bind the contents together. The cremaster, another substantial sheet of muscle, starts at the abdominal wall, extends down the cord, and also surrounds the testes (Fig. 4–16A and B). It aids in regulating the temperature of the testes, warming them by bringing them closer to the abdomen in the cold and helping to cool them by dropping them lower when it is hot. This is an important function because the testes require a temperature somewhat lower than that of the rest of the body for proper sperm production.

Some Common Problems of the Male Reproductive Organs

One of the consequences of the migration of the testicles from an internal location within the abdomen to the scrotum is that a break in the continuity of the muscle and fascia of the abdominal wall is formed, causing a weak spot in the wall. At either end of the inguinal canals are muscular rings, the **internal** and **external inguinal rings** (Fig. 4–16A). They are prone to expand when excessive pressure is exerted on the wall of the abdomen. This pressure can result from improper lifting of heavy objects, coughing, or any other form of abdominal strain. Too much or too frequent stretching of the rings can cause the canal to enlarge, even to the extent of permitting a piece of intestine to bulge into the canal and scrotum. Such a protrusion of abdominal contents into a region where it does not belong is called a **hernia** (rupture). A hernia in this region is called an inguinal hernia (Fig. 4–17). During a physical examination, the physician can diagnose this condition by placing a finger over the external ring and asking the person to cough. Even a weakening of this ring of muscle, which might predispose one to the development of a frank hernia, can be detected in this manner. (This is not a do-it-yourself procedure.) Surgery that repairs the injury and strengthens the muscles and fascia of the wall is the only remedy. Hernias can occur elsewhere (e.g., umbilicus and diaphragm), but the inguinal region is the most common site in men.

It does happen occasionally that one or both testes fail to descend totally. They may remain in the abdomen or become stuck in the inguinal canal. This condition is called **cryptorchidism.** Corrective surgery is usually recommended for the undescended testicle because the high temperature in this internal location inhibits sperm formation and because abdominal testes sometimes develop other problems. Failure of the testes to descend, however, does not affect other aspects of masculinity because the interstitial cells are not affected and do continue to secrete hormones.

Hydrocele is a collection of fluid between the layers of tissue that cover the testicles. Hydrocele is frequently found in newborns, another consequence of the complex migration of the testicles during gestation. It can also occur later in life after physical injury, infections of the epididymis, and gonorrhea. Minor surgery can correct hydrocele.

Varicocele is another common but minor problem. A varicosity is an abnormal dilation of a vein and can occur in many parts of the body. A varicocele is a dilation of the spermatic vein (pampiniform plexus). The condition

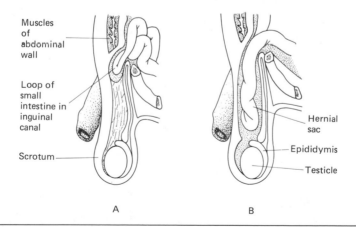

Muscles of abdominal wall

Loop of small intestine in inguinal canal

Scrotum

Hernial sac

Epididymis

Testicle

A

B

Figure 4–17 Inguinal hernias. *A,* Early stage. *B,* An advanced hernia with an intestinal loop extending into the scrotum. (After Arey: *Developmental Anatomy,* W.B. Saunders Co., Philadelphia, 1965.)

may be painless or may occasionally cause severe pain. It can also be corrected with minor surgery. This condition is sometimes associated with male infertility (see Chapter 18).

Epididymitis is an inflammation of the epididymis caused by infection or physical trauma. The suffix –**itis** means inflammation and, when added to the name of an organ, means inflammation of that organ. Epididymitis is often caused by a gonorrheal infection that has ascended the vas deferens. The symptoms are tenderness and swelling. It is treated by drugs, rest, and suspension of the scrotum to avoid pressure. Treatment is necessary to prevent epididymitis from developing into a chronic condition.

The most common problem of the male organs is **prostatitis.** It is usually induced by an infection but can also be caused by other factors. It is important to treat a prostatic infection vigorously to prevent it from spreading to other parts of the urinary system or from becoming a chronic condition. Prostatitis from any cause can be unpleasant because the gland surrounds the urethra (Figs. 4–10 and 4–12), and consequently the swelling that accompanies inflammation can result in a painful restriction of urine flow through the prostatic urethra. The physician can readily determine the size, shape, and texture of the prostate gland by inserting a finger into the rectum and palpating the gland (Fig. 4–10). Rectal palpation is routinely done in a medical examination because the prostate is a trouble spot in the male urogenital system.

Many portions of the male reproductive tract can develop cancer, but most of these, such as the penis and testicles, are relatively rare cancer sites. The exception is the prostate gland, which is prone to develop cancer. Nearly 16 percent of all cancers in men are prostatic, second in frequency only to lung cancer (21 percent). However, this form of cancer has a very low death rate because it occurs primarily in older men and is slow-growing and amenable to treatment. Diagnosis of prostatic cancer is made by rectal palpation.

CHAPTER 5

Anatomy and Physiology of the Female Reproductive Organs

The female reproductive organs are classified as *internal* and *external*. The internal female reproductive organs are the (1) ovaries, (2) oviducts, (3) uterus, and (4) vagina. The external female reproductive organs are the (1) mons veneris, (2) labia majora, (3) labia minora, (4) Bartholin's and Skene's glands, (5) clitoris, and (6) breasts.

The Internal Organs

The ovaries are discussed in Chapter 2. The **oviducts** are 4-inch (10-cm) tubes that interconnect the ovaries and the uterus. They are also known as fallopian tubes or simply tubes. Their main functions are to carry the egg to the uterus and, conversely, to allow sperm to reach the newly released egg, since fertilization normally occurs in the oviducts. Their terminal ends are open, expanded, and formed into finger-like folds (Figs. 5–1 and 5–2), which partially surround the ovary. At ovulation, these folds move over the surface of the ovary and totally envelop it, thus ensuring that the released egg is captured in the oviduct rather than allowed to slip into the abdominal cavity.

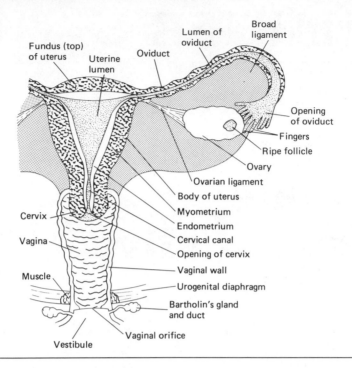

Figure 5–1 The female reproductive organs in diagrammatic frontal section.

The internal structure of the oviducts fits the generalized description of the typical organ shown in Figure 4–6. The lumen (hollow center) is covered by an internal epithelium that is formed into a complex pattern of folds (see Fig. 5–3A), so that the lumen is actually microscopic in dimension. These epithelial cells are bordered with hairlike cilia, which beat in waves that gently carry the egg toward the uterus (see Fig. 5–3B). However, sperm cells manage to buck this current and can reach the ovarian end of the oviducts within minutes after ejaculation. One of the mechanisms that promote this action is a spasmodic contraction of the smooth muscles of the uterus and oviducts that tends to propel the sperm toward the ovary. This contraction is stimulated by intercourse and by prostaglandins, chemical constituents of semen.

The main functions of the **uterus** (womb) are to nurture the mammalian embryo and fetus until birth and to expel it at that time. The uterus is often described as having the size and shape of a pear (Fig. 5–1). In side view, it is curved and rests on top of the bladder (Fig. 5–2). Any other position (retroverted uterus) can cause menstrual and fertility problems. The uterus is supported primarily by a large flat sheet of connective tissue, the **broad ligament,** which also connects it with the ovary and oviducts (Fig. 5–1). The uterus consists of a **body, fundus** (top), **cervix,** and **lumen.** The cervix is a necklike region that extends into the vagina. A short but important **cervi-**

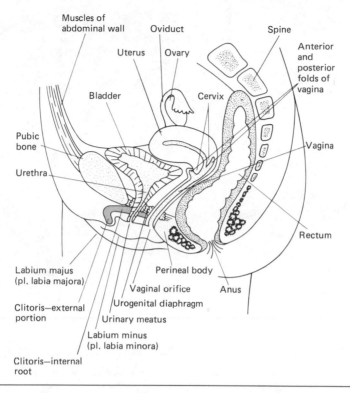

Figure 5–2 Organs of the female pelvis in a midsagittal section.

cal canal extends through the cervix into the vagina. This is the external orifice of the uterus.

The histology of the uterus fits the "typical" pattern shown in Figure 4–6. The smooth muscle layer of the uterus, the **myometrium** (Greek *myo,* muscle; *metris,* womb), consists of interlacing bundles of fibers rather than the typical arrangement of two layers at right angles. Consequently, when the uterine muscles contract, the uterus shortens simultaneously in all directions, somewhat analogous to making a fist with the hand. This action facilitates the expulsion of a baby at parturition. The inner lining of the uterus is called the **endometrium.** It consists of a layer of simple columnar epithelium and numerous tubular glands. These uterine glands extend deeply into a richly vascularized bed of connective tissue (Fig. 5–4). This endometrial layer undergoes conspicuous cyclical changes, which are described in detail in the next section.

The cervical canal also contains numerous glands (see Fig. 18–5). For the greater part of the menstrual cycle, these glands secrete a thick, viscous variety of mucus that effectively plugs the cervix and prevents the passage of sperm and pathogenic bacteria. At the time of ovulation, high concentrations of estrogen (see Chapter 8) stimulate the cervical glands to secrete copious

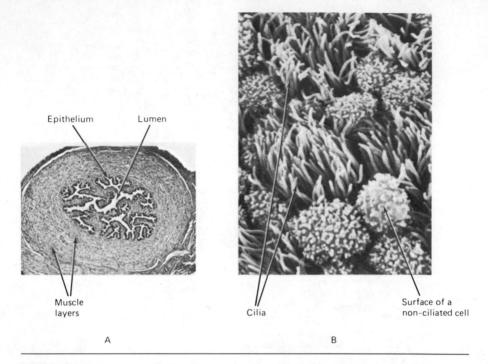

Epithelium   Lumen

Muscle
layers

Cilia

Surface of a
non-ciliated cell

A

B

Figure 5–3 Oviducts. *A,* Photomicrograph of a human oviduct in cross section. Magnification 15×. An egg would just about fit between the spaces. *B,* Surface view of oviduct epithelium at very high magnification, about 3000×, taken with a scanning electron microscope. The surfaces of only 10 cells are shown. Note the tufts of microscopic cilia, which beat in synchronous waves to propel the egg gently to the uterus. (From Bloom and Fawcett: *A Textbook of Histology,* 10th ed. W.B. Saunders Co., Philadelphia, 1975.)

amounts of a very thin mucus, which actually facilitates the entrance of sperm into the uterus by forming fluid channels in the cervix. These changes in the consistency of cervical mucus are sometimes used to determine ovulation time (see Chapter 18).

The **vagina** is a simple tube about 3 to 4 inches (7.5 to 10 cm) long (Figs. 5–1 and 5–2). The walls are normally collapsed but can expand to accommodate an erect penis during intercourse or the head of a baby at birth. The vaginal epithelium, like that of the skin, is composed of flat, stratified cells that make a fairly tough lining. This epithelium is surrounded by a layer of dense connective tissue. The internal end of the vagina does not terminate at the cervix but extends a little beyond (Fig. 5–2). This extension forms a circular fold of vaginal tissue around the cervix (a convenient platform for a contraceptive diaphragm). The vagina opens into the vestibule after passing through the urogenital diaphragm. There are no muscles that extend the length of the vagina, but some strands of muscle within the urogenital diaphragm surround the

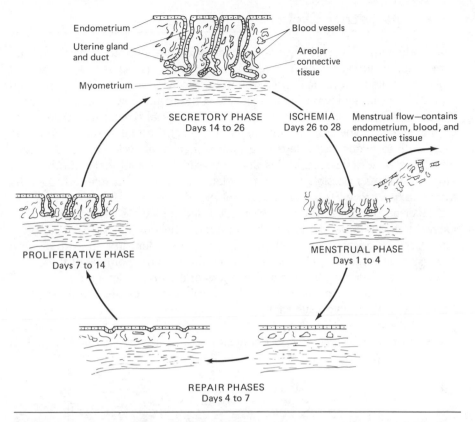

Figure 5—4 The human endometrial cycle.

orifice and can constrict it. The contraction of other muscles of the pelvic floor (e.g., the levator ani) can also constrict the vagina.

The **hymen** is a loose, perforated membrane that partially occludes the vaginal orifice in young girls. The vagina in the fetus is a blind pocket that is completely sealed by the hymen. Late in fetal life this membrane starts to break down. Occasionally, it may remain imperforate, causing problems when menstruation begins and the flow cannot escape. The hymen may be unusually tough in some women and may have to be dilated gradually or even cut surgically before intercourse is possible. First attempts at intercourse may be somewhat painful, not only because of a stretching of whatever hymenal remnants may still be present, but also because of the initial stretching of the vagina. Remnants of the hymen may persist in some women after initial intercourse. Many young women, though virgin, have no apparent hymen because of inherent developmental differences, athletic activity, or use of vaginal tampons. Contrary to popular myth, the presence or absence of a hymen does not constitute proof that a woman is or is not a virgin.

The Endometrial Cycle

The endometrial layer of the uterus undergoes pronounced cyclical changes every month during a woman's reproductive years. These changes consist of a gradual buildup of tissue in preparation for receiving a fertilized egg, followed by an abrupt breakdown of this layer and a menstrual flow if pregnancy does not occur (Fig. 5–4). The endometrial and superficial connective tissue layers of the uterine lining are at their maximum height, about 6 mm, during the fourteenth to twenty-sixth days of a typical menstrual cycle. The endometrial glands are well developed at this time and secrete a fluid that nourishes any embryo that may possibly be present. Hence, this period is referred to as the **secretory phase.** There are abundant blood vessels present in this superficial layer at this stage. They are elongated and follow a spiral course as they extend from the basal portion to the surface epithelium (Fig. 5–5). If no pregnancy occurs, an appropriate chemical signal (a sudden drop in estrogen and progesterone levels; see Chapter 8) makes the terminal portions of these vessels constrict, depriving the superficial layers of the uterine lining of a supply of oxygen. This lack of oxygen, called **ischemia,** then causes these tissues to degenerate. About two days after the vasoconstriction takes place, the affected parts of the uterine lining begin to be shed (Figs. 5–4 and 5–5). This is the menstrual flow, which consists of epithelium, connective tissue, and blood. This

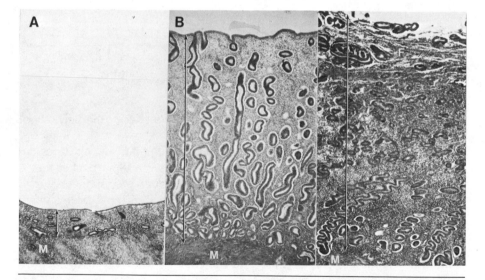

Figure 5–5 Photographs of uterine endometrium at different stages, all at the same magnification, 10×. *A,* Postmenstrual repair phase. *B,* Secretory stage. *C,* Early menstrual phase in which the upper layers are beginning to break down as a menstrual flow (F). Compare the thickness of the endometrial layer, which is indicated by two-headed arrows, at the different stages. A small portion of the myometrial layer (M) is present in these photos.

blood normally clots in the uterus, and the clots are broken up by enzyme action before flowing from the cervix. However, if menstruation at any given time is abnormally rapid, the blood may clot in the vagina. The normal blood loss during a menstrual period is variable but averages about 50 ml (2 oz). Menstruation takes about four days, with a normal range of three to six days.

The endometrium, now only about 1 mm thick, then goes into a three-day **repair phase** in which (1) continuity of the epithelium is re-established, (2) the remaining blood vessels start to grow upward again, and (3) gland development is initiated again as invaginations of epithelium form (see Chapter 4). Rapid growth of the epithelium, blood vessels, and connective tissue takes place during the next week, the **proliferative phase.** This is followed by another secretory phase. All of these changes are closely timed with those occurring in the maturing follicle of the ovary (see Chapter 2). These endometrial and follicular changes will all be correlated with the hormonal alterations discussed in Chapter 8.

Many other tissues of the reproductive tract also undergo changes during the menstrual cycle, though these changes are not as dramatic as those of the endometrium. Among these are changes (1) in the motility of the cilia of the oviduct (maximal at ovulation), (2) in the viscosity of the mucus that plugs the cervix (lowest at ovulation, to permit passage of sperm); and (3) in the cells of the vaginal epithelium. They can best be summarized as a series of elaborate and delicately integrated preparations to promote fertilization shortly after ovulation and to facilitate implantation several days later.

The External Organs

The anterior-most aspect of the external genitalia of the female is the **mons veneris** (mount of Venus, also known as the mons pubis), which is a pad of fatty tissue overlying the pubic bone. The **labia majora** (Latin *labi,* a lip) are thick, fatty folds that extend from the mons and enclose all of the other external genitalia (Fig. 5–6). The mons and the outside surface of the labia majora are covered with pubic hair. The labia terminate just above the anus in the **perineal body** (Fig. 5–6). This body is a central anchoring point for many muscles and fascial layers of the pelvic floor. The **labia minora** are a pair of smaller folds of tissue just within the labia majora. They are not normally visible externally but may protrude beyond the labia majora in some women. The **vulva** is the region between the labia majora. The **vestibule** is the area between the labia minora and contains the openings of the vagina, urethra, and two pairs of glands. The terms *vulva* and *vestibule* are sometimes used synonymously, but there is a slight difference in their definitions.

The **urethra** in women is very short, about 1½ inches (3½ cm) long. It extends from the bladder to the **urinary meatus** (opening) in the vestibule (Figs. 5–2 and 5–6). The meatus is an almost invisible opening about halfway between the clitoris and vaginal orifice.

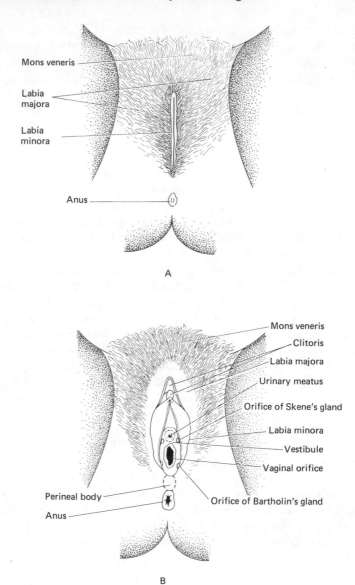

Figure 5–6 External genitalia of women. *A,* Natural appearance of genitalia. *B,* Appearance when the labia are separated.

The accessory sexual glands of women are those of **Skene** and **Bartholin.** Skene's glands are a pair of small glands that open adjacent to the urinary meatus (Fig. 5–6). The paired Bartholin's glands (Figs. 5–1 and 5–6) have openings adjacent to the vaginal orifice. The function of both Skene's and Bartholin's glands is to secrete a fluid to keep the vestibule moist.

The **clitoris** is located just above the labia minora (Fig. 5–6). It is homologous to the penis in structure and embryonic origin. Externally, it has a small shaft terminating in a glans, which is covered by a prepuce. Internally, it has deep roots in the pelvic floor (Fig. 5–2), which are similar to, though smaller than, those of the male penis. The roots contain spongy tissue and each is covered by an ischiocavernosus muscle. The clitoris can become erect during sexual excitement, but this does not occur in all women. Because it is richly supplied with sensory nerve endings, this organ is the principal site of sexual stimulation in women.

The **breasts,** though they are located on the anterior chest wall and are some distance from the other reproductive organs, are an integral part of this system. Externally, the conical breasts terminate in a nipple, which is surrounded by a pigmented area of uneven skin, the **areola** (Fig. 5–7A). The skin underneath the areola contains some wisps of smooth muscle that contract on mechanical stimulation, raising the nipple and facilitating nursing. Sexual excitement or cold will also stimulate this response. The breasts are composed of small islands of glandular epithelium interspersed with considerable amounts of fatty connective tissue. The function of this glandular tissue is to secrete milk for the newborn. The amount of glandular tissue is very small in the breast that is not actively lactating (forming milk) (Fig. 5–8). However, it increases considerably during pregnancy in preparation for lactation (see Chapter 15). Most of the secreting tissue is deep within the breast. It merges into a series of ducts, usually 15 to 24, each of which empties separately onto the surface of

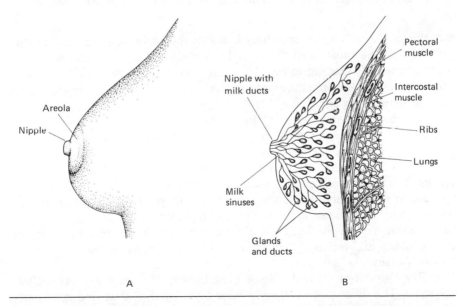

Areola

Nipple

Nipple with
milk ducts

Milk
sinuses

Glands
and ducts

Pectoral
muscle

Intercostal
muscle

Ribs

Lungs

A

B

Figure 5–7 Breast structure. *A,* Surface view of a female breast. *B,* Internal anatomy, lactating state.

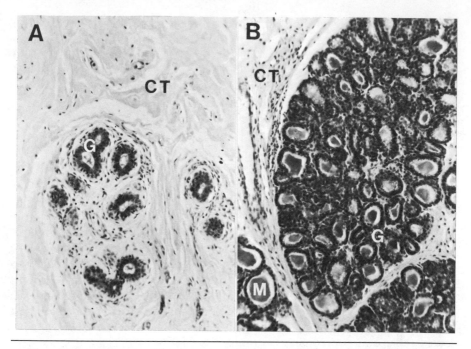

Figure 5–8 Microscopic structure of breast tissue. Magnification 20×. *A*, Resting stage. Note the small amount of glandular tissue (G) and the abundant connective tissue (CT). *B*, Lactating stage. The amount of glandular tissue (G) is greatly increased, and some glands are engorged with freshly formed milk (M).

the nipple. There is no skeletal muscle within the breast itself to provide support, though the muscles of the underlying chest wall (Fig. 5–7B) are sometimes erroneously thought to be part of the breasts. The only supporting tissue within the breast is connective tissue, which, like all connective tissue, slowly loses resiliency with age.

The breasts start to develop in the six-week embryo as the skin epithelium in the region of the nipple forms several ingrowths of glandular tissue (see Chapter 4). These mammary buds remain dormant until an early stage of puberty when the glandular tissue starts to grow and fatty deposition occurs. This change is usually accompanied by some tenderness. The glandular tissue of the breasts is stimulated to develop again in early pregnancy because of hormone action. This breast growth, also accompanied by tenderness, is a common early sign of pregnancy. Most of the enlargement of the breasts stimulated by pregnancy and lactation is lost afterward, but in many women, the breasts remain significantly larger.

The amount of glandular tissue in all breasts of mature women is about the same. The difference between a large breast and a small one is primarily due to the amount of fatty tissue present. Consequently, breast size has no relationship to the ability of a woman to breast-feed a baby.

Accessory or supernumerary breasts develop in about 0.1 percent of women. These always form along the milk lines, two conspicuous strands of tissue present in the young fetus that extend from the axilla, move down the chest and abdomen, and terminate in the inguinal region. Usually a supernumerary breast consists simply of a rudimentary nipple or appears merely like a pigmented mole. It is rare that a complete supernumerary breast forms.

Some Common Problems of the Female Reproductive Organs

The reproductive organs can frequently be a source of physical problems for a woman. The majority of these problems are minor, but some can be serious. Many of the potentially serious conditions can be detected early and can be prevented from developing to a dangerous state; therefore, it is important that all women more than 20 years of age have annual **gynecological** examinations (i.e., examinations of the female reproductive organs). In such an examination, the breasts and the pelvic organs specifically are inspected. To facilitate the examination, the woman lies on her back with her knees bent and her feet in stirrups. The external genitalia are checked for cysts or infection. An instrument called a speculum is placed into the vagina. It allows the physician to open up the walls of the vagina, which are normally collapsed. The vagina and the cervix can then be checked for abnormalities. If there is a vaginal discharge, a drop can be taken for microscopic examination. A few cells will be scraped gently from the cervix, smeared on a glass slide, and preserved for examination later for possible signs of early cancer. This is a **cytology** test, also known as a Pap smear, named for George Papanicolaou, who developed the technique. The speculum is removed, and the gynecologist performs a manual examination of the pelvic organs by placing two gloved fingers in the vagina and pressing down with the other hand on the abdomen. The size, shape, and consistency of the uterus can be determined. The ovaries can often be palpated in this manner also. This examination of the pelvic organs is sometimes supplemented by palpation through the rectum. In addition, the breasts are manually examined for tumors and other possible problems.

The two most common gynecological problems of young women are menstrual disorders and vaginal discharges associated with vulvar irritation. The latter disorder is called vaginitis, an inflammation of the vagina, and is usually caused by an infection. It is discussed in detail in Chapter 7. Problems of menstrual flow can be classified as flow that is unusually heavy or light, excessively short or prolonged, abnormally frequent, or totally absent. Uterine bleeding can also occur at some point between normal cycles (spotting). Most of these problems result from hormone imbalances, which can occur for a variety of reasons (see Chapter 8). Sometimes these difficulties may be caused by a physical disorder of the vagina or uterus or even by a systemic disease (e.g., blood clotting disorders and severe vitamin deficiencies). Cessation of

CAMROSE LUTHERAN COLLEGE
LIBRARY

menstruation for several months can be caused by pregnancy, hormone problems, or undernutrition (e.g., crash diets) or may occur after prolonged use of oral contraceptives (see Chapter 18). This condition is also frequently caused by psychological stresses such as fear of pregnancy, death in the family, or a move to a new locale. The vast majority of these menstrual problems can be corrected by hormone therapy, by treatment of a specific disorder in the uterus or vagina, or sometimes by time alone. Prolonged periods of vigorous exercise (e.g., marathon running and ballet training), especially when combined with stringent dieting, can also lead to chronic menstrual irregularity.

Toxic shock syndrome (TSS) is a rare but severe illness that can affect menstruating women. It is marked by a sudden onset of fever, vomiting, diarrhea, low blood pressure, and, in severe cases, shock. A minor epidemic of TSS occurred in the United States in 1980, when 870 cases were reported with a fatality rate of 8 percent. Even though the incidence was low (3 per 100,000 menstruating women at the peak of the epidemic), the dramatic nature of the symptoms attracted considerable attention. It was quickly traced to the use of superabsorbent tampons (which were then quickly taken off the market) and to normally innocuous bacteria that are common residents of the skin and vagina. The prolonged retention of superabsorbent tampons in the vagina apparently allows the bacteria to flourish and give off a toxin, which produces the illness. Awareness of the problem and the withdrawal of superabsorbent tampons reduced the incidence of TSS in 1981 to almost half the 1980 level. But this problem has not been eliminated, since it can also occur with the use of regular tampons. The risk of TSS for women who use tampons can be minimized if women use them only intermittently during a menstrual period. The risk can be reduced to virtually zero by not using tampons at all.

Cystitis is an inflammation of the bladder caused by infection or trauma. The chief symptoms are a frequent desire to urinate and a burning pain during urination. Cystitis can be precipitated by a period of frequent intercourse because of mild trauma to the urinary meatus, urethra, and bladder, all of which are very close to the vagina. This disorder is sufficiently common to have earned the name "honeymoon cystitis." Cystitis is treated with antibiotics and the consumption of large amounts of water to help flush the microorganisms out of the bladder.

Cysts are fluid-filled sacs and in the female external reproductive organs may range in size from microscopic to an inch or more in diameter. They are usually innocuous and painless by themselves, but they may cause trouble if they impinge on a sensitive structure or become infected. Cysts are prone to develop on the vulva, particularly in or around Bartholin's glands. Most of these can be ignored, but a large or infected cyst can be painful and is treated in the physician's office by incision, drainage, and a course of antibiotics. The ovary is also prone to the development of cysts, some of which may become quite large (up to 3 inches). The fact that these can become so large does not necessarily signify a serious condition. The cause is most often failure to ovu-

CAMROSE LUTHERAN COLLEGE
LIBRARY

late, and hormone therapy will often reduce even large ovarian cysts. Surgery is occasionally necessary to remove painful cysts.

Polyps are mushroom-like outgrowths (tumors) of the mucous membranes. They may occur in several areas of the body; the endometrium and cervix are common locations. Polyps are similar to cysts in that they are innocuous and can sometimes be ignored, but since uterine and cervical polyps often interfere with menstruation or cause midcycle bleeding, they are usually removed by a minor surgical procedure.

A **tumor** is an abnormal mass of tissue. If the mass remains well localized and does not invade other tissues, it is called a **benign tumor.** Tumors can often be ignored unless they cause pain or physically interfere with the function of an organ. A type of tumor that spreads by invading other tissues is much more dangerous and is called a **malignant tumor,** or **cancer.** It may be difficult or impossible to determine whether a tumor is benign or malignant by simple physical examination. A small piece of the tumor can be surgically removed for microscopic examination in a laboratory. This procedure is called a **biopsy** and almost always provides a definite diagnosis.

The female reproductive tract can develop both benign and malignant tumors. A common form of benign tumor that develops in the myometrium is known as a **fibroid tumor.** This is a hard mass of smooth muscle that is usually innocuous and decreases in size after menopause. Particularly large or numerous fibroid tumors may prevent pregnancy or cause excessive menstrual flow. In the latter case, surgical removal of the uterus, called **hysterectomy,** may be necessary. Cancer is not uncommon in the generative organs, since 22 percent of all malignant tumors in women occur in the breasts, 5 percent in the ovaries, and 17 percent in the uterus. That is the bad news. The good news is that all of these sites can easily be examined for tumors by direct inspection, palpation, or cytologic tests (Pap smear).

The dangerous and eventually lethal effects of cancer are best prevented by early diagnosis and treatment. Thus everyone should take advantage of the methods for early diagnosis that are available. The treatment for cancer is surgical removal of the growth, radiation therapy, or chemotherapy, or some combination of these. Radiation and anticancer chemicals kill cells that are dividing (in mitosis). Since malignant tumors generally consist of populations of rapidly dividing cells, they are differentially sensitive to these agents.

CHAPTER 6

Physiology of Intercourse

The primary function of the reproductive system is to reproduce, to produce individuals for the next generation. This process requires that gametes from two individuals of opposite sex come together and that the product of their union, the fertilized egg, be nurtured until it is capable of some degree of independent existence. Sperm are normally introduced into the female body during intercourse; the friction between the penis and vagina precipitates the ejaculation of sperm, which are deposited in the base of the vagina. The sperm can move from that location into the internal female organs to meet the egg and fertilize it. As far as reproduction is concerned, all that is required of intercourse is that sperm are deposited into the vagina. But intercourse cannot simply be equated with the first stage of the reproductive process. Sexual activity also encompasses many important emotional aspects, some of which are discussed in Chapter 19. Although the text of this chapter will be confined solely to the physiology of intercourse, the student should be aware that this is just one of the many facets of sexual activity.

The Four Stages of Sexual Response

Sexual activity has not been considered an appropriate area for scientific study until relatively recent times. The pioneering studies on the physiology of the human sexual response by the physician and psychologist team of William Masters and Virginia Johnson were initiated in the 1950s. Their findings were summarized in their 1966 book, *Human Sexual Response,* which not only added new dimensions to our knowledge of this field, but also helped to dispel many longstanding misconceptions.

Masters and Johnson divided the sexual response into four phases, which are comparable in men and women. They are (1) excitement, (2) plateau, (3) orgasm, and (4) resolution. The **excitement phase** is marked by the onset

62

of erotic feelings. The blood supply to the pelvis is increased (vasocongestion); the penis becomes erect in the man, and the vagina becomes lubricated in the woman. The **plateau stage** is a more advanced state of arousal occurring prior to orgasm. The **orgasm stage** is the height of pleasurable sexual sensations. It is characterized in both sexes by rhythmical contractions of the pelvic musculature. The **resolution stage** is marked by a return to normal, a reversal of the pelvic vasocongestion and general abatement of all sex-specific physiological responses.

The Male Sexual Response

The general pelvic vasocongestion of the excitement phase in men is associated with dilation of the central arteries of the corpora cavernosa of the penis, leading to an erection (p. 42). There is often a reddening of the skin, called the sex flush, which may extend from the abdomen to the face and neck, and an increase in the rate of the heart beat and breathing. The scrotum also thickens, and the testes are elevated because the spermatic cords become shorter.

Pelvic vasocongestion is at its peak during the plateau stage. There are minor increases in the size of the penis, especially in the glans penis.

The orgasmic phase has two components. The first occurs when the smooth muscle layers of the prostate gland, seminal vesicles, and vas deferens contract and start to deliver semen into the urethra. This is accompanied by a well-defined conscious feeling of ejaculatory inevitability. The delivery of semen into the urethra is followed by a series of involuntary rhythmical contractions of the bulbocavernosus muscle, which covers the base of the corpus cavernosum urethrae, and other muscles of the perineal region. These contractions, which occur at intervals of 0.8 second, propel the ejaculate through and out of the penis in spurts, as well as provide part of the orgasmic sensation.

After ejaculation, the veins in the penis open up and start to drain the excess volume of blood. This occurs in two stages: an initial rapid stage, which quickly reduces somewhat the size and turgidity of the penis, and a slower secondary stage, which eventually reduces the penis to its usual flaccid state. The rapidity with which either of these two stages occurs depends on the individual and the degree of sexual stimulation that may persist. Men enter a refractory period after orgasm, and some time must elapse before ejaculation can occur again. Other changes in this resolution phase are a reversal of pelvic vasocongestion and a return to normal heart rate and breathing.

The Female Sexual Response

There is a definite series of well-defined events in the female sexual response, even though none of them are as obvious as penile erection and ejaculation.

Sexual stimulation initiates vascular engorgement of the pelvic organs during the excitement phase, which in turn leads to (1) vaginal lubrication by an exudate from the vaginal walls, (2) swelling and enlargement of the uterus, and (3) some degree of swelling of the spongy tissue in the clitoris. The breasts may also swell, the nipples become erect, and the skin of the chest and face becomes flushed and mottled. An important effect of this excitement phase is a relaxation and enlargement of the vagina, which, along with the onset of lubrication, prepare the vagina for penetration by the penis. All of the changes in the excitement phase usually occur more slowly in the woman than in the man, in whom the erection is the primary occurrence of this initial stage. It is for this reason that foreplay, the gentle building up of a state of arousal by sexual stimulation, is even more important for the woman than the man. An attempt to enter a vagina that is not relaxed and lubricated can be uncomfortable, even painful, for the woman.

Pelvic vasocongestion increases during the plateau phase, which usually occurs during intercourse, and causes a swelling and deep reddening of the labia minora. The deeper portions of the vagina balloon out, and the tissues around the vaginal orifice become congested. Even though there are no muscles deep within the vagina, vaginal constriction during intercourse can be increased for the heightened pleasure of both partners by consciously contracting all the muscles of the pelvic floor, similar to the motion used to terminate urination. The clitoris is the principal location of sexually responsive nerve endings; therefore, the physical stimulation of this organ in some manner during the excitement and plateau phases is necessary to bring a woman to orgasm. The clitoris will retract to some extent just prior to orgasm.

The female orgasm consists of involuntary rhythmical contractions of the muscles around the vaginal orifice as well as other muscles and congested tissues of the pelvic floor. These contractions, like those of the male orgasm, occur at intervals of 0.8 second. Unlike the man, who enters a refractory period after orgasm, the woman frequently can be stimulated to repeated orgasms while she remains in the swollen plateau phase.

There is a rapid decongestion of the pelvis during the resolution stage. The deep reddening of the labia minora subsides and the clitoris returns to normal within seconds after orgasm. It may take 10 to 20 minutes, however, before the vagina and the uterus return to normal.

Dysfunctions of the Sexual Response

It is necessary to understand a few facts about the nerves that mediate sexual response before any discussion of sexual response problems can be effective. The physiological aspects of the sexual response are controlled by the autonomic nervous system, which is largely, but not exclusively, concerned with internal visceral functions such as the stimulation of gland secretion and the contraction of smooth muscle. The physiological changes brought about by this branch of the nervous system tend to be automatic and not under voluntary control. Thus, a man cannot voluntarily "will" himself an erection, though he might be able to induce one indirectly by having erotic thoughts. There are two divisions of the autonomic nervous system, the sympathetic and the parasympathetic. The latter division tends to control the internal physiological aspects of normal intestinal peristalsis and regular rates of heart beat and breathing. The parasympathetic division also mediates the physiological reactions of the excitement and plateau phases of the sexual response—that is, erection of the penis, pelvic vasocongestion, and vaginal lubrication. The sympathetic division prepares the body to handle stressful situations by increasing muscle tension, reducing peristalsis, and increasing the rates of heart beat and breathing. This division is readily stimulated, for example, by fear. The sympathetic division also mediates the physiological responses of the orgasm phase of sexual response. Organs controlled by the autonomic nervous system have innervations from both divisions, which have antagonistic effects on that organ. Consequently, while the sympathetic division is controlling an organ, the parasympathetic cannot.

The physiological responses of sexual interest and activity are mediated by the autonomic nervous system and thus are not ordinarily under voluntary control. Another important consequence of this type of control system is that since the bodily responses of the excitement phase, particularly penile erection and vaginal relaxation and lubrication, are mediated by the parasympathetic division, these events cannot occur or be maintained if the sympathetic division is switched on by any feelings of fear or concern. This simple fact is the psychophysiological basis for the two most common problems of sexual functioning, impotence and frigidity.

Impotence, also currently called erectile dysfunction, is the inability to achieve or maintain a penile erection in a sexual situation. Every man occasionally has a single episode or even a period of impotence. It is a problem only when the condition persists for an extended time. It can be caused by physical factors such as stress, fatigue, diabetes, narcotics, alcohol, some drugs (especially antihypertension medication), and neurological damage. More commonly, it is due to psychological factors—particularly those that introduce some degree of anxiety into the sexual experience—for the reasons outlined in the preceding paragraph. Sometimes merely a transient period of impotence leads to concern about sexual abilities, which has a cyclical effect in maintaining the impotence. This cycle is called performance anxiety. Therapy for impotence includes (1) eliminating possible physical causes, (2) eliminating anxieties that are interfering with the sexual response, and (3) performing exercises that promote the free flow of the sexual response.

Female sexual unresponsiveness, sometimes called frigidity, is a more complex situation. It may vary from simple inability to lubricate the vagina to total inability to have any sexual feelings or orgasm. An anxious impotent man is incapable of having intercourse, but an unresponsive female can still function as a sexual partner. However, if the fear or anxiety is considerable, an involuntary tightening of the muscles of the vaginal orifice (vaginismus) may make intercourse impossible. Therapy for any degree of sexual unresponsiveness in women is basically similar to that for men, namely (1) elimination of possible physical causes, (2) elimination of anxieties, and (3) exercises to develop the sexual response.

CHAPTER 7

Sexually Transmitted Diseases and Other Infectious Disorders of the Reproductive Organs

All openings of the body and all warm, moist areas are particularly susceptible to infectious disorders. Since the reproductive and urinary systems have both openings and moist areas, they are often infected by a variety of **pathogenic** (i.e., disease-producing) bacteria, viruses, and other parasitic organisms. Many of these organisms specifically infect only the tissues of the genital tract. Many of them also fail to thrive even for brief periods outside the body. The spread of such an infection, therefore, almost always requires direct contact with an infected genital tract. Such diseases are called **venereal** (Latin *venereus,* pertaining to Venus, the goddess of love). Another term widely used for these diseases at present is **sexually transmitted diseases,** or STD.

Syphilis and **gonorrhea** are the diseases most commonly thought of when venereal diseases are mentioned, but there are about 14 diseases classified as STDs because (1) they are much more prevalent in sexually promiscuous groups than in the general public, (2) they are virtually absent in celibate individuals, and (3) there is a high rate of simultaneous infection in sexual partners. Most of the disorders discussed in this chapter belong in the category of STDs because they meet all these criteria, although some (such as pubic lice and some types of vaginitis) can also be spread without relation to sexual activity.

Syphilis

Syphilis is caused by a bacterium-like organism called *Treponema pallidum*. These tiny organisms have a corkscrew shape and hence are often referred to as **spirochetes.** There is excellent evidence that this disease was introduced to Europe in 1493 by Columbus and his sailors after their first trip to the West Indies. Although this may never be established incontrovertibly, it is well documented that devastating epidemics of syphilis spread from Spain to Italy to France and then throughout Europe between 1495 and 1497, largely because of military operations. The Portuguese introduced syphilis to India in 1498, and shortly thereafter it spread to China and Japan. This bit of history illustrates the highly contagious nature of this disease.

The organism that causes syphilis has often affected the course of history, even though it is a frail creature that dies very rapidly outside the body. This is the reason that syphilis can be transmitted only by intimate contact, usually sexual. An untreated infection typically goes through the following stages:

1. Infection is transmitted by contact, especially of moist tissue. The most common sites of initial infection are the penis, scrotum, vulva, vagina, mouth, and rectum.
2. An incubation period of about three weeks follows during which there are no symptoms.
3. Next is the occurrence of a primary stage, marked by the appearance of a hard, round ulcer called a **chancre** at the site of the initial infection. A microscopic examination of a scraping from the chancre will show that it is teeming with spirochetes. The chancre is usually painless and heals spontaneously within a few weeks.
4. If the syphilis is untreated, a secondary stage occurs from several weeks to several months after the chancre heals. The most common feature of this stage is a red skin rash, which may or may not be accompanied by headache, fever, and some symptoms in a variety of organs. These are generally mild, nonspecific symptoms that are rarely associated with the actual cause—syphilis.
5. A latent period, free of overt symptoms, which may last for many years or even for life, then follows. It is during this period that colonies of the spirochetes may invade the blood vessels, bones, brain, spinal cord, and other tissues and establish colonies there.

6. In about 30 to 50 percent of untreated patients, a tertiary stage occurs in which symptoms of serious brain, spinal cord, or heart disorders may appear.

Syphilis may also be transmitted from a pregnant woman to her fetus through the placenta or to the baby via the vagina during birth. Children with such congenital syphilis may develop hearing and visual problems as well as deformed teeth (see Chapter 17).

Syphilis can be diagnosed readily and accurately by a blood test. The identification under a microscope of spirochetes from a bit of tissue scraped from a chancre or other lesion is also diagnostic.

The incidence of syphilis in the United States is currently on a rather high plateau. From a low point of 5000 new cases per year during the 1950s, there was a rapid increase during the 1960s and a leveling off at about 20,000 new cases per year. Between 1977 and 1980 there was a 33 percent increase of reported cases to 27,000 per year. This seems to be a real increase and not a statistical artifact. Its significance is not known.

Fortunately, syphilis in both primary and secondary stages can be cured rather simply because the spirochete is very sensitive to penicillin. Very often, a single injection of long-acting antibiotic is adequate treatment. The patient should be checked several times for a year afterward to be certain that some organisms have not escaped and entered the insidious latent period.

Gonorrhea

The organism that causes gonorrhea (also known as "clap") is the bacterium *Neisseria gonorrhoeae*. It has been known since ancient Chinese and Egyptian times. This organism lives only in moist (mucous) membrane tissues such as those found in the genitalia, throat, and rectum. The usual site of infection in men is the urethra, and from there it may spread to the prostate, epididymis, testes, bladder, and kidneys. The usual site of infection in women is the cervix, but the organism can also be harbored in the urethra and in the glands of Skene and Bartholin. An untreated infection in women is particularly prone to spread (in about 15 percent of cases) into the uterus and fallopian tubes and from there to other organs of the pelvis, producing serious pelvic inflammatory disease (PID). Gonorrhea can also be transmitted during oral or anal intercourse, thereby infecting the throat or rectum. The organism enters the blood stream in about 1 percent of patients and localizes in joints, particularly those of the knees, wrists, and hands, to produce a painful arthritis.

The initial symptom of gonorrhea in men is a yellowish discharge from the urethra, which appears about two to ten days after infection. This is accompanied by a burning sensation during urination. The infection may subside spontaneously after a few weeks. It may also spread to other male organs or persist in a relatively asymptomatic chronic state, which is still infectious. The primary symptom of pharyngeal gonorrhea, particularly common in homosexual men, is a sore throat and sometimes fever. Rectal gonorrhea may produce a discharge and some itching but may also persist as an asymptomatic condition. A yellowish vaginal discharge may be present in early stages of gonorrhea in women, but in the majority of cases there are no symptoms. The spread of the infection into the uterus and oviducts is most likely to occur during menstruation, particularly the first menstrual period after infection. Severe pelvic pain, abdominal tenderness, and fever are usually the result of the ascending infection. The spread of the infection to the oviducts in either the acute or the chronic form results in scarring of these tissues, producing a condition known as "blocked tubes" and infertility (see Chapter 18). Gonorrheal infections are a major factor in producing this common form of infertility.

Unlike for syphilis, there is no simple and convenient test for gonorrhea. The usual methods are microscopic examination of discharges for visual evidence of the organism and culturing samples of the discharge in special bacterial culture media. Gonorrhea, like syphilis, can be treated with antibiotics. However, this is not a do-it-yourself treatment, since care must be taken to ensure that all the bacteria are killed and that no state of chronic infection develops. A recent complication in the treatment of gonorrhea has been the rapid worldwide spread of a penicillin-resistant strain of this bacterium. This is a cause of great concern because, up to now, penicillin has been the most effective and economical antibiotic for this disease.

There is a current worldwide epidemic of gonorrhea of considerable proportion. In the United States alone, there are 21,000 new cases reported weekly, more than a million annually. The reported cases are estimated to be only about one third of the actual cases. The worldwide incidence of new cases is estimated to be more than 200 million per year. The great majority of these cases are in individuals 15 to 29 years of age. The factors that contribute to this epidemic are (1) the highly contagious nature of the disease, (2) the large number of asymptomatic carriers, and (3) the increasing frequency of heterosexual and homosexual promiscuity.

Minor STDs Caused by Bacteria

Syphilis and gonorrhea are considered the major venereal diseases because of their great frequency and the severity of the symptoms they produce. There are, however, other infections of the reproductive organs that can be transmitted by sexual contact.

Nonspecific urethritis is an infection of the urethra by an organism that has not yet been identified. The symptoms, a burning sensation during urination and a discharge, are similar to but less severe than those of gonorrhea. The treatment of choice is tetracycline for several days. It is more common in men than in women, but women may be asymptomatic carriers.

Chancroid is caused by bacteria. The ulcer that forms is similar to that of syphilis but is quite painful. Granuloma inguinale, also known as chronic ve-nereal sores, is caused by bacteria that produce painless, progressively spread-ing skin sores. Lymphogranuloma venereum is caused by an agent that is nei-ther a typical bacterium nor a virus but has characteristics of both. The chief symptoms are fever and enlarged, tender lymph glands of the groin. All three of these disorders are relatively uncommon in the United States and do re-spond to antibiotics.

Viral STDs

Herpes genitalis is caused by a variety (type 2) of the herpes simplex virus that causes cold sores on the lips (type 1). This virus is transmitted by sexual contact. It causes blisters over the infected site, usually the glans penis, vagina, and vulva. The blisters eventually break open and leave small but painful ul-cers, which usually disappear after several weeks. The symptoms can be par-ticularly distressing to women because the ulcers on the vaginal wall and vulva are very painful during urination. Herpes genitalis, like cold sores, tends to recur at intervals. The presence of active genital lesions in a woman at the time of vaginal delivery can result in a very serious infection of the newborn that can be prevented only by a cesarian section (see page 165). This condition appears to be very common, similar in magnitude to gonorrhea. Treatment by a physician is available but is not always satisfactory.

    Venereal warts (condyloma acuminatum) are caused by a virus and resemble warts on other parts of the body. The most common sites of occur-rence are the base of the penis, under the foreskin, the vulva, the perineal area, and around the anus. They appear from one to eight months after con-tact with an infected partner. The order of infectivity is low and may require predisposing conditions such as persistent moisture. Treatment is available but must be performed by a physician.

    Cytomegalovirus is a benign virus that spreads by sexual contact. The transmission rate is low, but it is estimated that most individuals (about 75 percent) acquire it sometime during their lives. It would probably be consid-ered just one of the numerous innocuous microorganisms that normally inhabit the skin and mucous membranes of the body, except that on rare occasions it is transmitted to a fetus during pregnancy, with serious consequences (see Chapter 17).

Vaginitis

The vagina normally harbors a population of bacteria that help to maintain a healthy state within this organ. Occasionally, intruding organisms take over and produce unpleasant discharges and irritations of the vagina and vulva. Inflammation of the vagina is one of the most common of gynecological problems, especially among young women. The most frequent causes of **vaginitis** are infections with a fungus *(Candida),* a protozoan *(Trichomonas),* or a bacterium *(Hemophilus).* The last two in particular can be transmitted by coitus and hence are STDs. The male carriers in these cases are usually asymptomatic. Vaginitis can also occur from chemical irritation, such as from strong douches and vaginal deodorants, and from mechanical irritations, such as from foreign objects and intercourse without sufficient lubrication, either natural or artificial. Excessive douching can also predispose a woman to vaginal infections.

Candidiasis is the most common cause of vaginitis. It is caused by infection with a fungus of the genus *Candida.* The main symptom is a burning and itching of the vulva. White patches of discharge can be seen in the vaginal walls on examination. The fungus is a common inhabitant of vaginal tissue and may exist without producing symptoms. Certain conditions, however, allow this organism to flourish. These include pregnancy, diabetes, the use of oral contraceptives, and a course of treatment with antibiotics, which apparently interferes with the normal bacterial check-and-balance system. The infection is treated by applying fungicidal preparations, either creams or suppositories, in the vagina for an extended period.

Trichomoniasis is caused by infection with a unicellular (protozoan) parasite. The chief symptoms are a green to gray discharge with an offensive odor and severe itching of the vagina, vulva, and sometimes the entire perineal area. Diagnosis is made by microscopic examination of the discharge for the parasite. Treatment with drugs is effective, but it must be extended to the sexual partner for maximum effectiveness, since men can be asymptomatic carriers and since it is not unusual for partners to pass the infection back and forth.

Vaginitis can also be caused by the bacterium *Hemophilus vaginalis,* which produces a gray discharge that is slightly acidic and has an unpleasant odor. The treatment is antibiotics, preferably given to both sexual partners.

Pubic Parasites

A case of the **"crabs"** is caused by an infestation with a small blood-sucking louse (insect) called *Phthirus pubis.* Since these **pubic lice** cannot survive more than a day off the body, transmission is usually by sexual contact but can also occur via bedding and nonsexual contact. **Scabies** is caused by infestation with the mite *Sarcoptes scabiei,* which burrows into the skin and eventually raises itchy, red lumps. This parasite can be transmitted by sexual contact, although it can also infest nongenital areas and be spread by nonsexual contact. Infestation by both of these parasites can be controlled by specific medicated lotions.

Some Generalizations

It is important to note that the incidence of syphilis and gonorrhea is very high, particularly in young people. Some experts estimate that nearly half of American youths will become infected with either of these before they are 25 years old. Most of the readers of this book should be aware that they are in this particular high-risk group. These major STDs and most of the less dangerous, less common types are transmitted only by sexual contact of some kind. Therefore, avoiding exposure to these diseases becomes a commonsense matter: either avoid sexual contact or know your partner. The reason that these problems are much less common in individuals over 30 years old is not that sexual activity or sensitivity to infection is decreased, but rather that more people in this age group have stable relationships, which obviously carry less risk of infection.

It is important to remember that both syphilis and gonorrhea can be very dangerous diseases, even though the individual may be asymptomatic for long periods of time. Fortunately, these diseases and most other STDs respond well to medical treatment, though self-treatment should not be attempted. It should also be obvious that both members of a sexual partnership should be treated even though only one shows symptoms, to avoid the Ping-Pong effect.

CHAPTER 8

Endocrine Control of the Reproductive System

General Characteristics of Hormones

Chemical signals are very important for both the development and the normal functioning of reproductive organs. These chemicals are **hormones,** potent substances that are produced in very small quantities in one part of the body and that exert an important effect on some other part. The various hormones and the glands that secrete them constitute the **endocrine system.** The main functions of this system are to control and integrate bodily functions, which it does in conjunction with the nervous system.

The glands that have been discussed earlier deliver their secretion to a specific area via a duct. **Endocrine glands** have no ducts and deliver their secretions directly into the blood stream, which quickly distributes them throughout the body. Another important characteristic of the potent chemicals synthesized by endocrine glands is that they usually have multiple effects on more than one organ. The sex hormones are a fine example of this feature. Their primary target organs are one or more parts of the reproductive system, but they also affect behavior, the general physiology of the whole body, and the differentiation of bones, muscles, and fat deposits.

Hormones interact with each other to form a delicately integrated system of checks and balances. The **pituitary,** an important endocrine gland, secretes a number of **trophic hormones** that control the secretory activities of other endocrine glands. After the hormones produced by the stimulated gland reach a high concentration in the blood, the hormones themselves cause the pituitary to lower its secretion of trophic hormone and thus slow down the activity of the target organ. This type of negative **feedback control** is very important in the control of sex hormone production and is discussed in detail later in this chapter.

Some Hormones that Indirectly Affect the Reproductive System

The **thyroid gland** is found in front of and alongside the larynx (voice box). The hormone that it produces is **thyroxine,** the only hormone that contains the element iodine. The general function of this hormone is to control the rate of metabolism of the body. Thyroid deficiency causes a decrease in oxygen consumption, a reduction in the rate of heart beat, and slower reactions of the nervous system. This general sluggishness is often associated with a marked lack of **libido.** Thyroid deficiency can also adversely affect female fertility. An excess of thyroid hormone causes an increase in the utilization of oxygen and food, nervousness, and increased activity. The pituitary gland controls the thyroid by secreting a **thyrotrophic hormone.** Inadequate iodine in the diet can also affect thyroid hormone secretion.

The **pancreas** is a gland with a dual function. It synthesizes many digestive enzymes that are delivered directly to the intestine via the pancreatic duct. It also has a ductless endocrine component, which secretes the hormone **insulin.** This hormone controls the availability of sugar to cells. Insulin deficiency results in the well-known disease **diabetes.** This disease is characterized by an inability of sugar to get into some cells where it is used in the production of energy. Diabetes can affect pregnancy and the functioning of male reproductive organs.

The **adrenal glands** are located on top of the kidneys. Each has a central medulla that secretes hormones related to nervous system activity. The adrenal **cortex,** the outer layer, secretes many hormones: **Cortisone** and **aldosterone** are two that affect the body's reactions to stress and help control kidney function. The adrenal cortex is also important because it is a secondary source of sex hormones and secretes some estrogens and androgens in both sexes. The functioning of this gland is controlled by **corticotrophin,** which is secreted by the pituitary gland.

The Pituitary and the Hypothalamus

The **pituitary gland** is sometimes called the master endocrine gland because it controls so many other glands. Located in a bony pocket in the floor of the

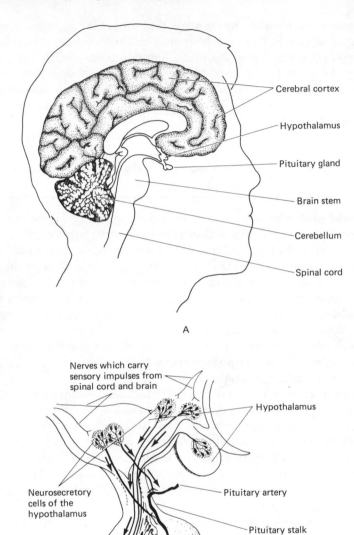

Cerebral cortex

Hypothalamus

Pituitary gland

Brain stem

Cerebellum

Spinal cord

A

Nerves which carry
sensory impulses from
spinal cord and brain

Hypothalamus

Neurosecretory
cells of the
hypothalamus

Pituitary artery

Pituitary stalk

Posterior
lobe of
pituitary

Anterior lobe
of pituitary

B

Figure 8–1 The hypothalamus and the pituitary gland. *A,* Relationship of these structures to the brain and head. *B,* Enlarged view showing the manner in which the neurosecretory cells of the hypothalamus relate to the pituitary. The sensory input into the hypothalamus from other parts of the nervous system is also diagrammatically indicated.

skull, just above the roof of the mouth, it is a globular organ, about ½ inch (12 mm) in diameter and is attached directly to the brain by a stalk (Fig. 8–1). It is divided into posterior and anterior lobes.

Two of the hormones associated with the posterior lobe are **antidiuretic hormone** and **oxytocin.** The former prevents diuresis (copious urination). Oxytocin causes the contraction of smooth muscle. It functions in returning the uterus to normal size following birth. Oxytocin is sometimes given to induce labor because it will cause the uterine muscles to contract. This hormone also stimulates the milk ejection reflex in lactation (p. 162).

The anterior lobe secretes numerous important hormones. **Growth hormone** controls growth in youths. Too much can cause gigantism; an inadequate amount can lead to the development of a midget. **Prolactin** stimulates lactation. The anterior lobe also secretes trophic hormones such as thyrotrophin, corticotrophin, and gonadotrophins, which stimulate the thyroid, adrenals, and gonads, respectively, to secrete their hormones.

The **pituitary gonadotrophins** control the endocrine functions of testis and ovary. There are two of them, **luteinizing hormone (LH)** and **follicle-stimulating hormone (FSH),** and both are produced in both sexes. LH stimulates the development of the corpus luteum in women and controls hormone secretion in men by stimulating the interstitial cells of the testes. LH is sometimes given an alternate name in men: **interstitial cell—stimulating hormone (ICSH).** The other gonadotrophin, FSH, initiates ovarian follicle development in women and stimulates sperm production in men. These trophic and other hormones of the pituitary gland exert a profound effect on all parts of the body, an effect that is disproportionate to its small size.

In turn, the pituitary gland is controlled not only by the very glands that it stimulates through the feedback effects of the hormones they secrete but also by a part of the brain, the **hypothalamus.** This is a relatively small area located above the stalk of the pituitary. It directly controls the posterior lobe by nerves that extend between the two regions. The control of the anterior lobe is more complex. Some of the nerve cells (neurosecretory cells) of the hypothalamus secrete hormones, called **releasing factors,** which are carried directly to the anterior pituitary by a unique miniature system of blood vessels (Fig. 8–1). There are several releasing factors, one for each pituitary hormone controlled. **FSH releasing factor (FSH-RF)** is carried by these blood vessels to the cells of the pituitary that synthesize FSH and there controls their activity. **LH releasing factor (LH-RF)** similarly is carried to and controls the activity of the cells that synthesize LH.

The hypothalamus is like a highway for many important pathways of nerves within the brain. It receives sensory information from all parts of the body as well as stimuli from other parts of the brain. This nerve input affects the functioning of the hypothalamus in its role as a control center for the pituitary gland. Thus the hypothalamus integrates the two coordinating systems of the body, the endocrine and nervous systems.

The Male Sex Hormones

The principal male sex hormone is **testosterone,** a term often used inclusively for all male sex hormones. This is a convenient but not strictly accurate usage, since there are several closely related compounds in the body as well as some synthetic hormones that have male hormone activity. These are collectively known as **androgens.**

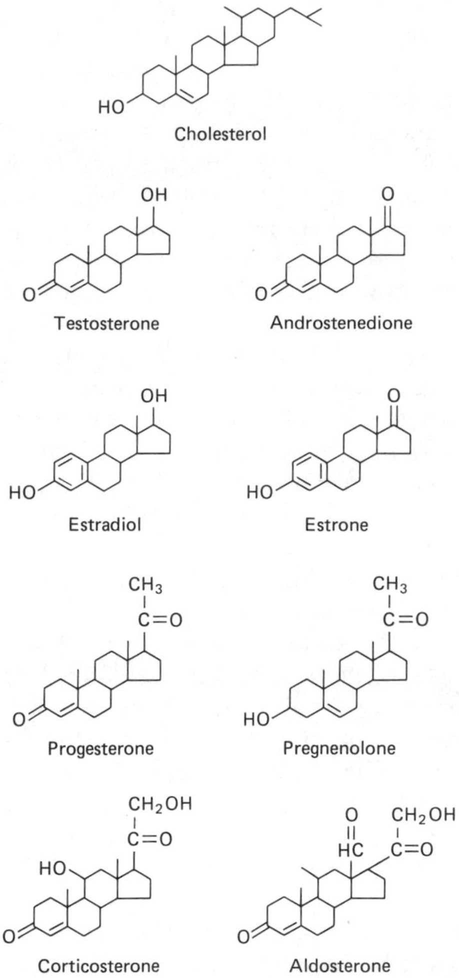

Figure 8–2   The chemical structure of some steroid compounds. Testosterone, estradiol, and progesterone are the major sex hormones. Androstenedione, estrone, and pregnenolone are closely related to them. Corticosterone and aldosterone are adrenal cortical hormones. Cholesterol is the parent molecule for the synthesis of all of these. Very small differences in chemical structure can result in great differences in biological activity.

The molecular structure of testosterone and other androgens is based on a chemical unit called **sterol** (Fig. 8–2). This unit is also the basic portion of all estrogens and progesterone as well as cholesterol and some hormones of the adrenal cortex. The synthesis in the body of all of these important molecules is interrelated, a significant feature of reproductive endocrinology.

Most of the testosterone in men is synthesized by the interstitial cells of the testes (Fig. 4–9), but very small quantities are formed in the adrenal cortex in both men and women.

The chief functions of the androgens are to (1) initiate and control the prenatal and pubertal development of all parts of the male reproductive system, (2) maintain the male sexual organs in a healthy and vigorous state, and (3) stimulate libido in both men and women. The androgens also have a variety of effects on many organs and tissues that are not immediately related to reproductive functioning and therefore are called **secondary sex characteristics.** The presence of testosterone in the blood stream during puberty and on a continuous basis after that influences the following:

1. Hair pattern, stimulating the beard and body hair in general and, paradoxically, inducing baldness (see Chapter 16).
2. Fat distribution, which is generally sparser in men; belly fat in men tends to accumulate above the umbilicus, whereas it accumulates below the waist in women.
3. Skin texture and color; in men the skin is generally darker and not as smooth as in women.
4. Skeleton and muscles, generally stimulating them to develop more robustly in men than in women.
5. Metabolism; the greater oxygen consumption and rate of protein synthesis and breakdown in men is testosterone-induced.
6. Voice, which tends to be deeper in men because the laryngeal cartilages that form the voice box are stimulated during puberty to develop in a larger and thicker fashion than in women; the vocal cords are also stimulated to develop into thicker strands, making the voice harsher.
7. Personality; hormone-induced courtship and display behavior is conspicuous in many animals and is less so, but not totally absent, in humans.

Control of Testosterone Production

The initial trigger for testosterone production, whether at puberty or at any other time, is the secretion by the hypothalamus of luteinizing hormone releasing factor (LH-RF). This stimulates the anterior pituitary to synthesize and secrete luteinizing hormone, which is also called the interstitial cell–stimulating hormone (ICSH). This agent then triggers the interstitial cells of the testes to synthesize testosterone and other androgens. The testosterone is released into the blood stream, from which it is selectively picked up by the target organs of

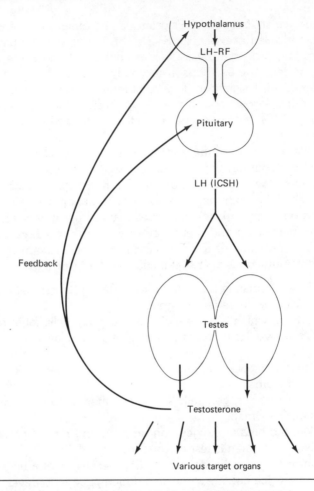

Figure 8–3 The interactions between the hypothalamus, pituitary, and testes, which constitute the feedback control mechanism for maintaining testosterone synthesis at a steady level.

the reproductive system (Fig. 8–3). Some of it reaches the pituitary and hypothalamus, which are sensitive to this hormone. The LH-secreting cells are turned off when blood levels of testosterone exceed a certain concentration, which, in turn, lowers the production of testosterone. Conversely, when the blood level of testosterone gets too low, the LH-secreting cells of the pituitary are switched on to secrete more LH, and the interstitial cells are stimulated again. This is a negative feedback control system. The net effect is that of a check-and-balance system that provides for the maintenance of a fairly steady level of testosterone at all times. This is a beautiful example of **homeostasis,** a general word for the mechanisms that keep hormone levels, blood sugar, body temperature, and many other bodily functions at a steady state level.

The negative feedback control system for testosterone works very well and is not easily upset. However, overloading the system with externally supplied testosterone, which some athletes do in order to build up their muscles, for example, can temporarily disrupt it. It may take several weeks to re-establish normal control after the hormone treatment is discontinued. Paradoxically, high testosterone levels produce temporary sterility because they inhibit the comparable FSH feedback mechanism, and FSH is necessary for sperm formation.

The Female Sex Hormones

There are two female sex hormones: **estrogens** and **progesterone.** Both share, with each other and with androgens, the same basic steroid structure (Fig. 8–2).

Several estrogens are synthesized in the ovary, particularly by the follicle cells. Their secretion is controlled by the follicle-stimulating hormone (FSH) of the pituitary and the FSH-RF of the hypothalamus. Their concentration varies with the stage of the menstrual cycle, being highest at the time of ovulation. Estrogens are primarily responsible for the maintenance of the female reproductive tract and the breasts. They are also responsible for the development and maintenance of the secondary female sex characteristics. A combination of the presence of estrogen and absence of androgen produces the following effects:

1. The body hair is sparser because the hair follicles, equally numerous in both sexes, produce short, downy hair instead of long, thick hair.
2. More fat is deposited on the hips, in the breasts, and under the skin.
3. Muscle and bone do not develop as robustly as in men.
4. The voice is generally lighter and higher pitched than in men because the development of the larynx and vocal cords at puberty is different.

Progesterone is synthesized chiefly by the corpus luteum. Its main function is to maintain the uterus during pregnancy, but it also has a part to play in each menstrual cycle. Progesterone synthesis is controlled by the luteinizing hormone (LH) of the pituitary and the LH-RF of the hypothalamus.

The minute quantities of androgens that are synthesized by the adrenal cortex of women must be noted because they have an important role in stimulating libido.

The Female Endocrine Cycle: Its Control and Relationship to Ovarian and Endometrial Cycles

Female hormone production involves two sex hormones, two gonadotrophins, and two hypothalamic releasing factors. The female hormones are not maintained at any constant level. They rise and fall in concentration, each in its own rhythmic pattern that is directly related to both ovarian and endometrial cycles. These factors contribute to a control system for the sex hormones that is much more complex than the simple steady-state system of testosterone control in men.

The first day of menstruation is considered day 1 of a new cycle because it is a clear, well-defined point. A 28-day cycle is used as a convenient average standard, even though cycles of 23 to 36 days fall into the range of normal.

The initial trigger for a new cycle comes from the hypothalamus, which secretes FSH-RF, which in turn stimulates the pituitary to synthesize FSH. The blood levels of FSH then rise over a two-day period, plateau for several days, show a small peak at ovulation, and taper off during the rest of the cycle (Fig. 8–4). The initial rise in FSH stimulates one of the several young follicles waiting in reserve to mature. As the follicle matures within a two-week period, the follicle cells secrete estrogen into the growing follicular cavity, as well as into the blood stream. Blood levels of estrogen rise slowly and are responsible for, among other things, the repair and proliferative phases of the endometrial cycle (Chapter 5). The stored estrogen is released along with the egg at ovulation, so that a peak level is reached at this time. Estrogen is maintained at a fairly high concentration until day 25. Then, if pregnancy has not occurred, the estrogen level drops precipitously (Fig. 8–4). This decrease in estrogen level is one of the major factors in precipitating constriction of the endometrial arteries with consequent ischemia, which leads to the onset of another menstrual period. It seems reasonable to assume that this drop in estrogen level is also a feedback stimulus to the hypothalamus and pituitary to initiate another cycle, but this has not been unequivocably proved. The factors that trigger a new cycle are still not completely understood.

Luteinizing hormone is secreted by the pituitary in response to hypothalamic LH-RF. The blood levels of LH reach a marked peak at midcycle, 16 to 24 hours before ovulation (Fig. 8–4). This high concentration of LH has two functions. One is to help trigger egg release, and the other is to stimulate the cells of the empty follicle to start forming a corpus luteum. The versatile follicle cells now start to secrete progesterone as well as estrogen. The blood level of progesterone gradually increases as the corpus luteum develops and peaks at about 22 days into the cycle. It drops precipitously if no pregnancy occurs because the corpus luteum starts to degenerate on about day 24. This decrease in progesterone level also helps to initiate menstruation. Should an embryo become implanted into the surface of the uterus on about the twenty-first day of the cycle, a chemical signal, a gonadotrophin formed by the embryo (Chapter 15), is sent back to the ovary. The corpus luteum starts a renewed burst of growth instead of degenerating and continues to secrete large amounts of both progesterone and estrogen. This corpus luteum of pregnancy and the hormones it secretes play a vital role in the maintenance of pregnancy. The continuous secretion of high levels of female hormones by this organ also turns off the cyclical hypothalamic-pituitary control system, which gradually becomes re-established after the pregnancy is over.

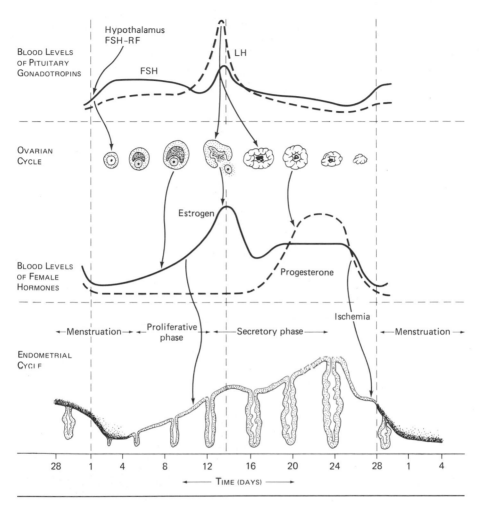

BLOOD LEVELS
OF PITUITARY
GONADOTROPINS

Hypothalamus
FSH-RF

LH

FSH

OVARIAN
CYCLE

Estrogen

BLOOD LEVELS
OF FEMALE
HORMONES

Progesterone

Ischemia

←Menstruation→ ←Proliferative
phase
←Secretory phase→
←Menstruation→

ENDOMETRIAL
CYCLE

28   1   4   8   12   16   20   24   28   1   4

←— TIME (DAYS) —→

Figure 8–4   The female endocrine cycle. Cause and effect interactions are indicated by the arrows.

This is a somewhat simplified version of a complex pattern of interacting events. Almost every statement of cause and effect mentioned in the preceding paragraphs could be qualified by the clause "but more than one factor is involved." It is not surprising that the female endocrine cycle can be disturbed to some degree by illness or any physical or emotional stress. It can be significantly altered by hormone administration because the hypothalamic-pituitary feedback control system is affected. The "Pill," which is a combination of estrogen and progesterone taken in a cyclical fashion, superimposes an artificially high level of hormones that mimics pregnancy, fools the hypothalamus, and thus prevents ovulation (Chapter 18).

Pregnancy drastically alters the pattern of the female cycle because of the temporary development of a different system of endocrine control, one that also involves several additional trophic hormones synthesized by the placenta (Chapter 15). Preparation of the breasts during pregnancy for secreting milk and control of this lactation also involve several hormones. This example of endocrine interaction is discussed in Chapter 15.

The female endocrine cycle is a beautiful example of a check-and-balance system of hormones, trophic stimuli, and brain impulses that control and integrate a complex pattern of cyclical structural changes of many parts of the body. Similar systems control the 28-day cycle of women, the 4-day cycle of rats, and the annual or semiannual cycle of some other animals.

CHAPTER 9

Puberty, Adolescence, and Menopause

Puberty

This chapter deals with the beginning and the termination of the reproductive capacity of the individual. **Puberty** (Latin *pubescere*, to be covered with hair) is the period when the reproductive system begins its final maturation. The end of puberty occurs when the individual is capable of reproduction. **Adolescence** (Latin *adolescere*, to grow up) is culturally defined as the stage between childhood and adulthood but is sometimes used synonymously with puberty. Puberty is marked by (1) structural and functional changes in the gonads; (2) onset of secretion of sex hormones; (3) development of functional interactions between the hypothalamus, pituitary, and gonads; (4) maturation of all reproductive organs; (5) development of the secondary sex characteristics; and (6) a final spurt of growth in height, at the end of which the long bones of the arms and legs are normally incapable of further elongation.

Puberty is initiated when the hypothalamus starts to secrete LH-RF and FSH-RF. These releasing factors then stimulate the pituitary to secrete gonadotrophins, which, in turn, initiate hormone secretion by the gonads. The establishment of the full system of stimulus, response, and feedback control develops slowly, even somewhat erratically, over several years.

Since the hypothalamus initiates puberty and the brain controls the hypothalamus, this is the route by which environmental factors can affect puberty. Both the onset and the course of puberty are known to be influenced by a number of factors. Puberty starts one or two years earlier in girls than in boys. Geography, genetics, and the general state of nutrition and health of the individual are all influential on the onset. The average age of menarche, or the onset of menstruation, in young girls of Moscow, for example, is 13 years; in certain rural groups in Russia, the age is 15 years. An interesting but unexplained statistic is a steady decrease in the average age of menarche in girls of several European countries and the United States over the past 120 years. It has decreased from about 16.5 years in 1850 to 13.5 years in 1970. A comparable change in the age of puberty has occurred in boys, since the records of boys' choirs indicate that the age at which boys' voices change has decreased over the last 100 years.

Puberty in Girls

The progress of puberty is somewhat more accurately studied in girls than in boys because menarche is a clear, well-defined criterion. The average age of menarche in the United States at present is 13, with a range of 10 to 16 years considered within normal limits. Menarche is preceded by changes that begin at about age 9 or 10. The ovaries start to enlarge, form some young follicles, and begin to secrete estrogens. The breasts enlarge as the dormant glandular epithelium starts to grow and fat begins to accumulate between the gland tissue. Some straight pubic hair appears, and the adolescent height spurt begins (Fig. 9–1). Kinky pubic hair forms between the ages of 10 and 12. The growth spurt progresses at a maximal rate. Skin pigmentation, a process also under the control of a hormone from the pituitary gland, becomes more apparent in the vulva and the areolae of the breasts.

Menstruation begins at about age 13. The first cycles are irregular and do not normally result in the release of ova. Changes in the pelvis (see Chapter 4) start to occur. Between ages 13 and 14, axillary hair appears, the pubic hair becomes dense, and breast enlargement is significant. The sweat and oil glands

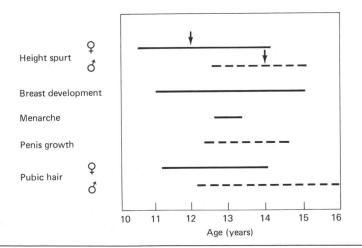

Figure 9–1 Indices of puberty in girls (solid bars) and boys (dotted bars). The bars indicate average periods of development; considerable individual variation is quite normal. The arrows indicate peak periods of growth in height.

of the skin enlarge and become susceptible to acne. Menstrual cycles often become regular by age 14. These are accompanied by ovulation and corpus luteum formation. Pregnancy can occur at this stage but is not recommended because it is accompanied by many more problems for both the mother and the child than are pregnancies of more mature women.

The preceding narrative concentrates mostly on the externally visible changes of puberty, but these are also accompanied by numerous developments in internal organs. The uterus, for example, increases tenfold in size between the ages of 10 and 15 years.

Puberty in Boys

Puberty in boys usually starts at about 11 to 12 years of age. The testes begin to grow. The seminiferous tubules, solid and dormant up to this age, start to become hollow and complete their differentiation. A few straight pubic hairs appear at this time (Fig. 9–1). Between the ages of 12 and 13, the interstitial cells of the testes increase in number and start to synthesize significant amounts of testosterone. A growth spurt of the penis begins. The pubic hair increases in quantity and becomes kinky. The adolescent growth spurt is at its maximum between the ages of 13 and 14 years. The voice begins to deepen in an erratic fashion as the larynx enlarges and the vocal cords thicken. Ejaculation is now possible as the prostate gland and seminal vesicles mature and start to secrete fluids. Sperm first appear in semen sometime between the ages of 15 and 17, and during this same time axillary hair appears, sweat and oil glands enlarge, and body fat decreases. The testosterone-stimulated development of muscles progresses rapidly at this time.

These are the physical indices by which the transformation of boys and girls into men and women is measured. Puberty is also an emotionally turbulent time. Although not all of the emotional changes characteristic of puberty can be attributed to the erratic surge of hormones in the blood stream, these upheavals are certainly strongly influenced by endocrine changes.

Menopause

Menopause is the period of reproductive senescence in women. Its onset and progress are generally measured by menstrual periods, because they start to become irregular, the time between them is longer, and the flow is progressively scantier. This is the usual pattern, although in some women menstruation may cease abruptly. Menopause is considered to be finished about one year after the last menstrual period. Menopause begins before age 45 in about 25 percent of women, between ages 45 and 50 in 50 percent, and after age 50 in 25 percent. The general state of health and other environmental factors can affect the time of onset of menopause.

Menopause is caused by a regression of the ovaries. There is a progressive decrease in the number of follicles and therefore progressively less estrogen. Surgical removal of the ovaries precipitates immediate menopause. The natural trigger for ovarian senescence is not known. It is definitely not the hypothalamic-pituitary route, since the pituitary gland responds to the lack of estrogen in a characteristic feedback style by producing more and more pituitary gonadotrophins in a futile effort to stimulate the ovary.

The physical symptoms of menopause are primarily those produced by a lack of estrogen. Most of the effects of estrogen induced during puberty are slowly reversed. The uterus becomes smaller, the vaginal wall becomes slightly thinner, and the glandular tissue of the breasts regresses (though the overall size of these may not change significantly). Abrupt changes in the amount of blood flowing to the skin can induce a flushing of the skin (hot flashes) and periods of profuse sweating (night sweats). Menopause is a period of adaptation to a change in body chemistry and, as such, can be accompanied by fatigue, dizziness, headaches, and insomnia. Since these symptoms are largely caused by estrogen deficiency, they can be alleviated to some degree by taking synthetic estrogens, although this may not be medically advisable for some women.

The natural levels of estrogen never diminish to zero even if both ovaries are removed because the adrenal cortex continues to secrete small quantities of both male and female hormones. Perhaps for this reason many women have an increased desire for sexual activity after menopause, although this is also influenced by the fact that a fear of pregnancy has been eliminated.

There is no comparable physiological cessation of reproductive capability in men. There is a slow decline in testosterone levels in men during their 60s and 70s, and a gradual decrease in their sperm count. Neither is enough to inhibit sexual activity or fertility.

UNIT TWO

Development of the Human Body

CHAPTER 10

Beginnings: The First Three Weeks of Life

In Chapter 1 the concept of development as an epigenetic process (new formation) was introduced. How do organs form from previously unformed material? When does the body take shape? All of us can visualize ourselves as newborns, with or without looking at the family photo album. We can perceive ourselves as miniature adults at this time, somewhat different in body proportion (large heads, small pelvises, and so forth) but still very human-like. Without much difficulty we can regress in time and further picture ourselves in the womb as progressively smaller "humanoids." But there comes a time when we realize we must have looked very different, that the fertilized egg, a simple spherical cell only 0.004 inch (0.01 mm) in diameter, had to undergo considerable transformation to become a small, human-like individual of several thousand cells. Numerous events must take place, both before and after fertilization, before this epigenetic transformation of a fertilized egg into a tiny new individual can take place. Therefore, the story of development begins at these earliest stages.

Development Before Fertilization

The moment of fertilization is the time at which the onset of development has traditionally been marked. But the sperm and the egg cell that have come into union have had interesting life histories predating this moment. The sperm cell was transformed from an ordinary-looking cell to one with a condensed nucleus and motile tail several weeks before fertilization (see Chapter 2). The history of the ovum dates back even further, for all the egg cells that a woman will ever have are already present in her ovaries three to four months before she is born. Thus, in a very literal sense, our own individual histories go back to a time before our mothers were born. The 20 to 30 years a human egg spends in the ovary is not an idle time. It is a critical period during which many preparations are made for the tremendous task that only an egg cell can do—namely, form an individual of the next generation. The obvious element that an egg must store is a certain amount of food, enough to last until implantation into the walls of the uterus occurs, a full week after fertilization. A less obvious feature of the development of an egg within the ovary is that numerous and complex changes take place in both the nucleus and the cytoplasm of the egg, which enable it in turn to develop after fertilization. The sperm can make only a limited contribution to these biochemical preparations for development because of its limited size. The egg has no such size constraints. It even has the assistance of the follicle cells, which intimately surround the egg and literally inject a variety of vital materials into it as it grows and matures. These biochemical preparations for development are synthesized in a dormant form and then stored until they are needed.

Another important aspect of prefertilization development shared by sperm and egg is the formation of special molecular configurations on their cell surfaces for attachment and subsequent fusion. Normal cells do not fuse when they come into contact. The special adaptations that develop on the cell surfaces of the sperm and egg are such that, as soon as they come into contact, the membranes rupture and the two cells become one.

Fertilization

Fertilization is that dramatic moment when the single ovum that is released from the ovary that month combines with 1 of 100 million sperm to start a new life. It is not simply a time when the sperm and ovum become attached, but also an involved process that takes several hours, an exacting procedure with stringent requirements. The failure rate is high.

The time requirements for fertilization are quite strict. Sperm must be present in the female reproductive tract for a minimal time, estimated to be about 7 hours, before one can successfully fertilize an egg. During this time, some important changes occur on the sperm head. This final phase of sperm maturation is called **capacitation**—the development of the capacity to fertilize. Sperm also lose their ability to fertilize within a day or two after being released into the female organs. Estimates of the life span of an egg vary from as little as 1 hour after ovulation to a maximum of 24 hours. Its optimal period of fertilizability is probably not much more than a few hours after ovulation. Both sperm and ovum are literally dying cells, and they must come in contact with each other before their deaths occur. These time considerations are significant for couples with fertility problems (see Chapter 18).

The tiny sperm must penetrate a formidable barrier of cells, called the corona radiata, and the egg membranes before it can reach the surface of the egg to fertilize it (Fig. 10–1). This penetration is the function of that specialized

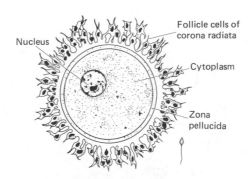

Figure 10–1   A drawing of a freshly ovulated human egg. The cells of the corona radiata surround and protect the egg but also constitute a considerable obstacle for the sperm. A human sperm cell is drawn to the same scale at the lower right. (Adapted from Arey: *Developmental Anatomy,* W.B. Saunders Co., Philadelphia, 1965.)

region of the head of the sperm, the **acrosome.** As the sperm approaches the cells of the corona radiata, the acrosome starts to break down and release some enzymes that digest intercellular material (Fig. 10–2). A narrow path forms between a few corona cells, and the sperm swims by. The last of the acrosome enzymes digests a path in the zona pellucida (Fig. 10–2C). The sperm head becomes attached obliquely to the egg, the cell membranes surrounding the egg and sperm fuse, and immediately the two cells are one (Fig. 10–2D). The contents of the sperm migrate into the substance of the egg. The condensed nucleus of the sperm starts to swell and then moves to the center of the egg, where it waits until the female nucleus finishes its maturation (Fig. 10–3). This male contribution is called a **pronucleus** at this stage because it contains only half of the normal genetic complement.

Rapid metabolic changes in the egg are initiated by sperm contact. A change in the surface membrane of the egg quickly blocks the attachment of any other sperm that might be close to the egg. Very rapid increases in protein synthesis and oxygen consumption begin to take place within seconds after fertilization. These changes can occur quickly because the mechanism for them has existed within the egg, albeit in an inactive form. Collectively, these metabolic changes are called **activation,** the awakening of the dormant biochemical machinery in the egg. This activation response, sparked by the sperm, is an important phase in sperm-egg interaction.

The female nucleus moves to the periphery of the egg after the entry of the sperm, where it then finishes its maturational divisions (see Chapter 3) and gives off the polar bodies. When this is finished, the female nucleus is also called a pronucleus. It then moves to the center of the egg, and the two pronuclei, one from the male parent and the other from the female parent, fuse as one (Fig. 10–3). This nucleus now contains the full complement of 46 chromosomes and is a new combination of genetic material. At this point, fertiliza-

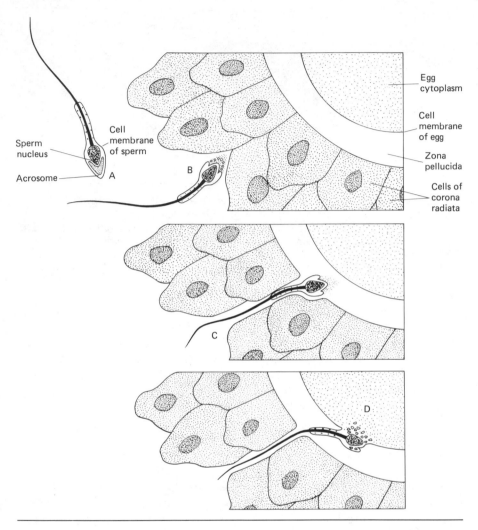

Figure 10–2 The fertilization of a mammalian egg (based on studies of laboratory mammals). *A,* The sperm approaches the egg. *B,* As the sperm touches the cells of the corona, the acrosome starts to break down and release some of its digestive enzymes. *C,* A path forms between the corona cells, and the sperm swims through and starts to form a path through the zona pellucida as the last of the acrosomic enzymes are released. *D,* Sperm and egg unite as the cell membranes of both fuse and become continuous. The size of the sperm is exaggerated somewhat to show detail.

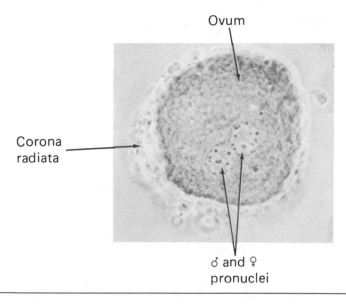

Ovum

Corona
radiata

♂ and ♀
pronuclei

Figure 10–3 The final phase of fertilization. A remarkable photograph of a human egg, which shows male and female pronuclei at the point of fusion into a single nucleus. (From O'Rahilly: *Developmental Stages in Human Embryos.* Carnegie Institution of Washington, Washington, D.C., 1973. Courtesy of Z. Dickmann and the *Anatomical Record, 152:293–302, 1965.*)

tion is considered finished, a leisurely 10 hours after the first contact of egg and sperm.

Fertilization is a complex, exacting process with many obstacles—chemical, physical, and chronological—that hinder the vital interaction between egg and sperm and significantly affect the odds for successful fertilization. The odds are not insurmountable, as the world's birth rate demonstrates, but these obstacles become important considerations for couples who are trying to overcome a fertility problem (see Chapter 18).

Cleavage and Implantation

Body organs are composed of tissues, and tissues are composed of cells. The period of development during which organs are formed epigenetically must be preceded by tissue formation, and tissue formation, in turn, cannot occur until a population of cells is present. Tissue and cell formation is an important preliminary of development. It is the primary feature of the first two weeks of development, along with implantation into the surface of the uterus and the establishment of a nutritional supply for the developing embryo.

The first three days in the life of the new individual are spent traveling through the oviducts, and four additional days are spent floating freely in the

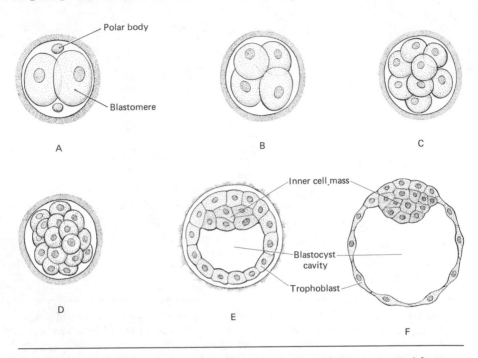

Figure 10–4 The first week of development: cleavage of the zygote and formation of the blastocyst. *A,* Two-cell stage. *B,* Four-cell stage. *C,* Eight-cell stage. *D,* Many-celled mass called a morula (Latin, "mulberry"). *E,* Early blastocyst, with cavity starting to form. *F,* Blastocyst just prior to implantation. (From Moore: *The Developing Human.* W.B. Saunders Co., Philadelphia, 1977.)

uterus, because implantation does not occur until a week after fertilization. During this period the fertilized egg, now called a **zygote,** divides into 2 cells, then 4 cells, 8 cells, 16 cells, and so on (Figs. 10–4*A, B,* and *C* and 18–1), until a lobulated, round mass forms (Fig. 10–4*D*). Since very little growth or change in the size or shape of the embryo has occurred, this period is called **cleavage,** a segmentation of the original mass of the egg into smaller cells. Once there is a population of about 50 cells, a fluid-filled cavity starts to form within the mass, and the embryo expands into a ball about $\frac{1}{16}$ inch (3 mm) in diameter (Fig. 10–4*E* and *F*). This ball, called a **blastocyst,** consists of a very thin-walled area called the **trophoblast** and a clump of cells at one side called the **inner cell mass.** The inner cell mass will eventually start to form the embryo, but not for still another week. The trophoblast consists of cells that will attach to the uterus and form a placenta.

    Implantation occurs on about day 7 or 8 as the cells of the uterine epithelium become receptive (Chapter 5) and the outer layer of cells of the blastocyst, the trophoblast, suddenly develops a "sticky" surface. The freely floating blastocyst then becomes attached to some part of the uterus (Fig. 10–

5A), and, within minutes, rapid changes start to occur in the cells at the point of contact. The cells that line the uterine wall move aside, letting the embryo come in contact with the underlying connective tissue. The cells of the tropho-blastic layer that come in contact with the uterine wall start to proliferate and invade the tissues of the uterus (Fig. 1C–5B). This burrowing takes place over a five-day period, at the end of which the embryo is completely surrounded by uterine tissue (Fig. 10–5C), in close proximity to many maternal blood vessels. A placenta as such has not yet formed, but simple diffusion is still sufficient to supply all the food and oxygen the tiny embryo needs. The prelim-inary stages of forming a population of cells and developing a nutritional sup-ply for them have been accomplished.

Formation of the Basic Tissues and Some Important Membranes

Internal rearrangements within the inner cell mass take place even as implan-tation is occurring. A cavity forms within the cell mass (Fig. 10–5B). This fluid-filled space is the amniotic cavity, and the membrane that surrounds it is the amnion. This is the "bag of waters" that will eventually surround the embry-onic body once it develops. Underneath this first ball of cells, a thin layer of cells appears, spreads, and forms another hollow ball (Figs. 10–5B and C and 10–6). This is the **yolk sac.** Fourteen days after fertilization, the blastocyst complex consists of a large outer ball within which are suspended two other balls in contact with each other. If you can visualize two water-filled balloons pressing upon each other, you can see that the interface between them would be a flat, two-layered disc. Such a flat, oval area forms between the amnion and the yolk sac. It is called the **germinal disc** because this is where the body of the embryo will soon start to develop.

The germinal disc at first has two sheets of tissue, an upper layer and a lower layer. Some internal rearrangements of cells occur within the germinal disc, and a middle layer of cells forms between the other two (Figs. 10–5C and 10–6). The formation of a three-layered embryo is an important stage because the basic body-building tissues are now present and organ formation can begin. These basic embryonic tissues are (1) **ectoderm,** the upper layer, which will form the skin and nervous system; (2) **endoderm,** the lower layer, which will form the epithelial lining of the stomach and intestines, as well as of a number of other internal organs such as the lungs, liver, and pancreas; and (3) **mesoderm,** the middle layer, which will form all the muscle and connec-tive tissue of the body, as well as some other internal organs such as the heart, blood vessels, kidneys, and genital system. A supply of both cells and tissues is now present in the germinal disc, and nutrition is guaranteed by the attach-ment of the embryo to the uterus. These thin sheets of tissue of microscopic dimensions are now ready to be converted into organs.

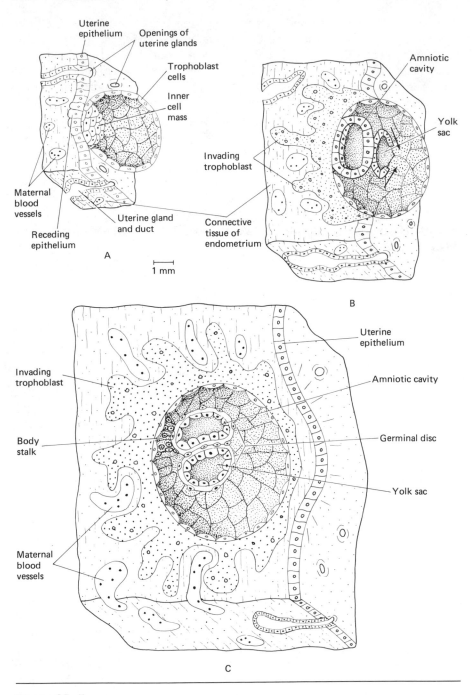

Figure 10–5

Figure 10–5 The second week of development. Three stages of implantation. *A,* Seven days after fertilization the blastocyst becomes attached to the uterine wall. *B,* About ten days. The noncellular trophoblast is invading the connective tissue of the endometrial layer and coming into close contact with maternal blood vessels. About half of the blastocyst is embedded. Meanwhile, an amniotic cavity has developed within the inner cell mass, and the yolk sac is starting to take a spherical shape. *C,* About 12 days. The blastocyst is completely embedded, and the uterine epithelium has covered it. The complex that will develop into the embryo is attached to the trophoblast (which will form the placenta) by a stalk of cells. Compare with the photograph in Figure 10–6.

Formation of a Primitive Body

By the end of the second week after fertilization, the three basic embryonic tissues are present as three flat layers in the germinal disc. These layers of cells are now sculptured into **primordia,** or beginnings, of organs by a variety of form-building movements (Fig. 10–7). Some organs are formed initially when a sheet of tissue simply **folds** to create a hollow groove or tube. Many organs

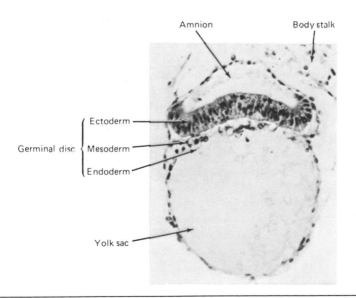

Figure 10–6 A photograph of a human embryo about 14 days old, found in the uterus of a woman who had a hysterectomy. The germinal disc is of microscopic size, 0.25 mm in diameter (0.01 inch) and 0.04 mm thick (0.0016 inch). Note the delicate, fragile, and relatively featureless state of the embryo a full two weeks after fertilization. (From O'Rahilly: *Developmental Stages in Human Embryos.* Carnegie Institution of Washington, Washington, D.C., 1973. Courtesy of Heuser, Rock, and Hertig.)

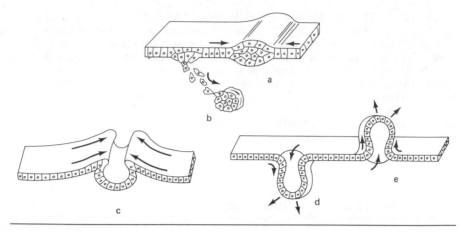

Figure 10—7 Form-building movements in a simple sheet of cells that lead to the formation of the primordia of organs. Cells can form clumps or rods *(a)* when they move toward a center or divide more rapidly in one place relative to another. Sometimes cells split off from a sheet, migrate, and reaggregate in another location *(b).* In *(c)* a sheet of cells is folding into a groove, the first stage in the development of many tubular organs. Many organs form from small pockets of cells that can extend either inward (an invagination) *(d)* or outward (an evagination) *(e).* The directions of cell movements are indicated by arrows. All organs of the body are formed initially from such simple changes in a small number of cells.

originate when groups of cells form microscopic pockets. Some cells **aggregate** into clusters or rods, which then form organs.

The first phase of body formation, which occurs during the third week after fertilization, is the development of a primitive body axis, a group of simple midline organs arranged in a short, straight line in the center of the germinal disc. The first structure to appear has a transitory but important history. It is an aggregation of cells in the midline of the posterior half of the disc known as the **primitive streak** (Fig. 10—8A). The streak terminates in the center of the germinal disc in a cluster of cells called the **primitive node.** Both node and streak play a vital role in the development of the primitive body. They do this by starting to move in a backward direction over a two-week period. As these structures retreat, the node causes rapid changes in all three of the primitive layers of tissues, influencing the cells that surround it to form the first organs of the body. Since the node moves backward in a straight line, these organs form in a line in front of the node. Because of its profound influence on this process of building the primitive body, the retreating node is sometimes referred to as an organizer center.

The first organs to form from the middle layer (mesoderm) are a **notochord,** a rodlike structure that eventually disappears, and a double row of aggregations of mesodermal cells, the **somites,** which flank the notochord and will eventually form the vertebrae (backbone). The notochord first forms

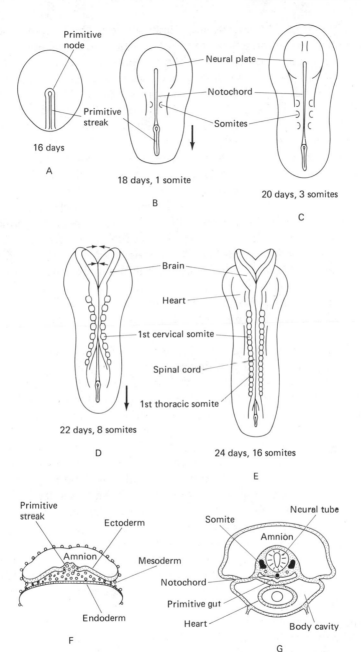

Figure 10–8 Development of the primitive body axis in human embryos during the third week. Surface views are shown in *A* to *E*. A cross section of a primitive streak stage embryo is shown in *F*, and one through the body of a 22-day embryo is shown in *G*. These illustrate the increase in complexity of the anatomy of the embryo.

immediately in front of the retreating primitive node. The node also stimulates the cells of the mesoderm adjacent to the notochord to aggregate into somites. These are formed in sequential fashion, one pair at a time in front of the node. Since the somites and additional lengths of notochord are added in a progressive fashion, the older an embryo is, the more somite pairs and axial notochord it will have. This can be seen in Figure 10–8. A 20-day-old embryo has 3 somites; 2 days later there are 8; at 24 days there are 16 pairs. The last of these to form is the pair directly in front of the retreating node.

The notochord temporarily provides some support and rigidity to the embryo and serves as a central axis for the primitive body. But its importance transcends these functions because it also sets into motion events that lead to the development of the brain and spinal cord. The notochord stimulates the cells directly above it, in the uppermost (ectodermal) layer of the germinal disc, to thicken and start to fold. The lateral edges of the thickened area rise above the flat embryo, first form a shallow "V" in cross section and then a deep "V," as shown in Figure 10–7C, and finally meet in the midline and fuse. After the edges fuse, they separate from the parent layer and form a free tube above the notochord. Thus a hollow tube forms from a flat layer. This is the humble beginning of the primitive **neural tube,** which will develop into the brain and spinal cord.

Here are the events of the third week of development. The primitive node moves backward and lays down a progressively longer notochord. The notochord stimulates the formation of a neural tube from the tissue above it. Since the neural tube develops above the notochord, which itself forms sequentially over a two-week period, it is obvious that the length of the brain and spinal cord also must develop in a progressive fashion. This can be more clearly visualized by counting somites again. The first four pairs of somites lie underneath the brain. They do not form vertebrae and will disappear. The fifth somite pair forms the first **cervical** (neck) vertebrae. It and the next six somite pairs that form mark off the seven cervical vertebrae of the neck. The neural tube that lies above these somite pairs is the cervical spinal cord. When the twelfth somite pair forms, it marks the start of the **thoracic** (chest) vertebrae and the thoracic spinal cord. The entire body axis is formed in this fashion, segment by segment, in head-to-toe sequence, until the fortieth somite pair in the tip of the tail is formed, about two weeks after the first somites. The central nervous system also develops bit by bit: first the brain, next the cervical spinal cord, then the thoracic spinal cord, and so on. It may be difficult to conceive of a human body developing in this manner, by progressively adding vertebrae and spinal cord from head to tail, but that is exactly what occurs.

As these events are occurring in the upper and middle layers of tissues, a simple midline tube of intestine called the **primitive gut** develops, also in a head-to-toe pattern, by the folding of the bottom layer of cells of the germinal disc. A tube of cells forms in the mesoderm by yet another movement of the basic tissues and develops into a primitive heart tube. The heart is the first organ actually to become functional. It starts to pulsate at the end of the third week. The first beats are feeble and erratic, barely more than twitches. They become rhythmical and a few days later propel blood, which carries nourishment and oxygen to all parts of the rapidly developing embryo via a primitive circulatory system.

The preceding paragraphs have described the manner in which body formation starts during the third week after fertilization as a primitive body pattern emerges. The embryo is about ⅛ inch long now, and the tissues of the germinal disc have undergone some transformations into (1) a microscopic brain and spinal cord in the midline; (2) groups of cells that flank the spinal cord and mark off the future vertebrae of the body; (3) a small, beating heart; and (4) a thin intestinal tube. This is a mere shadow of the magnificent body structure that is yet to form. Once this basic framework is present, many additional organs can, and do, develop within a short period of time, that is, the next three to four weeks.

Twins and Twinning

Although multiple births are common in many mammals, they are an interesting variation from the normal in humans. They have occurred with predictable frequency until this past decade. The relationship between the different kinds of multiple births was called the Rule of 89, an empirical relationship that fits the statistics. Human twins occurred approximately once in every 89 births. Triplets occurred once in 89^2 births, or approximately once in 8000 births. Quadruplets occurred once in 89^3 births, or approximately once in 705,000 births, and quintuplets in 89^4 births, or once in 63,500,000 births.

There are two kinds of twins. The more common type is the **fraternal,** or **dizygotic,** twin. Zygote refers to the fertilized egg; therefore, dizygotic twins are those who developed from two eggs that were simultaneously released and fertilized. Such fraternal twins can be of opposite sex and do not necessarily resemble each other more than any other brother or sister. There are genetic tendencies toward fraternal twinning in some families and in some groups of people, which simply means that some women have a tendency to release more than one egg at a time. This tendency has been greatly exaggerated recently by the use of fertility-enhancing drugs (see Chapter 18), which stimulate ovulation and frequently stimulate more than one follicle to ripen simultaneously. This has resulted in a rash of multiple human births of the dizygotic variety, making the Rule of 89 invalid at present.

The other type of twins is the **identical,** or **monozygotic,** variety. Monozygotic twins are individuals who arose in some fashion from a single cell. Consequently, these twins have identical chromosomes; they are always of the same sex and blood group and have the same bodily features. The frequency of monozygotic twinning is approximately equal in all groups of people. Identical twins result from accidents to eggs, accidents that can occur at several stages of very early development. The simplest fashion in which they

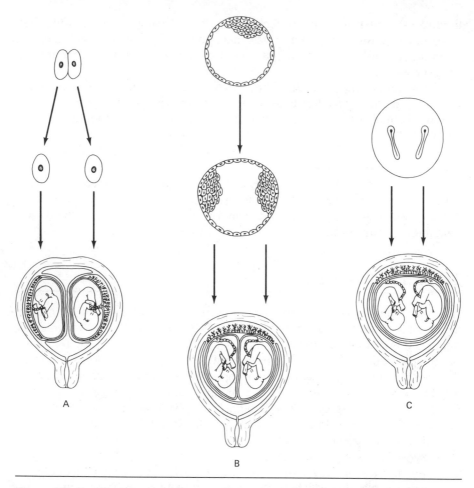

Figure 10–9 The origins of human identical twins by *A*, separation of cells at the two-cell stage; *B*, the development of two inner cell masses within a single blastocyst, probably the most common mode of origin; and *C*, the formation of two organizer centers within a single germinal disc. The uterine drawings (adapted from Arey: *Developmental Anatomy*, W.B. Saunders Co., Philadelphia, 1965) show different degrees of sharing of placentas and amnions. Generally speaking there is more sharing of membranes if the twins originate late.

can arise is by the separation of the two cells that develop at the earliest cleavage stages (Fig. 10–9A). Each cell then continues to develop independently into an embryo. This type of twinning can be readily produced in the laboratory with eggs of various invertebrates, frogs, and even mammals. These experiments demonstrate that, though the egg is normally the only cell that can form a new individual, this capacity is shared by the first few cells that develop from a fertilized egg. Another possibility is that somehow two inner cell masses arise within a blastocyst at about seven or eight days after fertilization, and each of these develops into an embryo (Fig. 10–9B). Since the node organizer center is responsible for the control of body formation, an accidental duplication of this organizer center can also lead to multiple births (Fig. 10–9C). This is precisely what happens in armadillos, which always have litters of four despite the fact that only a single egg is released. Four primitive streaks and nodes form at the blastocyst stage, and four identical armadillo quadruplets form in this fashion from a single egg. Some cases of human twins are also thought to occur in this fashion.

Identical twins may sometimes develop in such close proximity that they may never form totally separate bodies. Such twins are **conjoint,** that is, incompletely separated. More commonly they are called **Siamese** twins, named after Chang and Eng, the pair who were immortalized by P. T. Barnum.

CHAPTER 11

Development of the Body

At the end of the third week, the embryo is about 3 mm (⅛ inch) long. It is composed of many cells, some basic tissues, and a primitive body pattern laid down in a linear axis (Figs. 10–8D and 11–1). This simple arrangement of rods and tubes does not last very long as the existing structures start to change their shapes and many new organs come into being. Primitive tissue layers continue to fold into new shapes and new organs during the next five weeks. Groups of cells migrate and sometimes reaggregate at different locations to start new organs. The simple tubes that formed earlier now bend and twist and bud off microscopic pockets of cells, which also begin to develop into new organs. The second month of development is a period of rapid organ proliferation, a very dynamic time in the life history of a person. By the end of the second month, the new individual is about 1 inch (25 mm) long and resembles a miniature human for the first time. Internally, it contains the beginnings of virtually every organ it will ever have.

External Development of the Body

The oldest embryo pictured in the last chapter was 24 days old and had 16 pairs of somites (Fig. 10–8E). These somite pairs represent the neck levels and part of the chest region. The body-building activities continue during the fourth week. The node organizer center continues to lay down progressively the somites that will develop into the vertebrae of the lower body and the comparable levels of the spinal cord and other axial organs. Before the primitive streak disappears completely, it lays down 40 pairs of somites. The last few of these belong to the tail, a structure that is conspicuous in the young embryo (Fig. 11–2D and F) but that eventually disappears. Occasionally an individual is

106

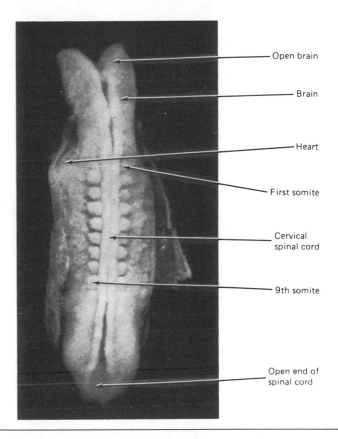

Open brain

Brain

Heart

First somite

Cervical
spinal cord

9th somite

Open end of
spinal cord

Figure 11–1 Photograph of a human embryo, 22 days old. Nine somite pairs
are present. The neural tube is still open at both ends. Note the straight-line axial
arrangement of all structures at this stage of development. (From Moore: *The Developing Human*. W.B. Saunders Co., Philadelphia, 1977.)

born with a small tail remnant that persisted instead of being resorbed. It can
easily be removed with minor surgery.

The young embryo of two to three weeks is flat (Figs. 10–8F and G and
11–3A), since the primitive body forms in the middle of a flat sheet—the germinal disc. The adult trunk (chest and abdomen) is essentially a cylindrical tube
stuffed with internal organs. This cylinder forms by a process of folding and
fusion (Fig. 11–3), similar to that which formed the brain. Primitive body cavities form to either side of the neural tube during the germinal disc stage as
spaces develop within the middle (mesodermal) layer. The flat tissues on both
sides of the primitive axial organs fold around, meet underneath the embryo,
and fuse together. Consequently, a rounded body is formed, and the two lateral spaces are brought together to create the body cavity.

During these movements the lowermost sheet of embryonic tissue, the

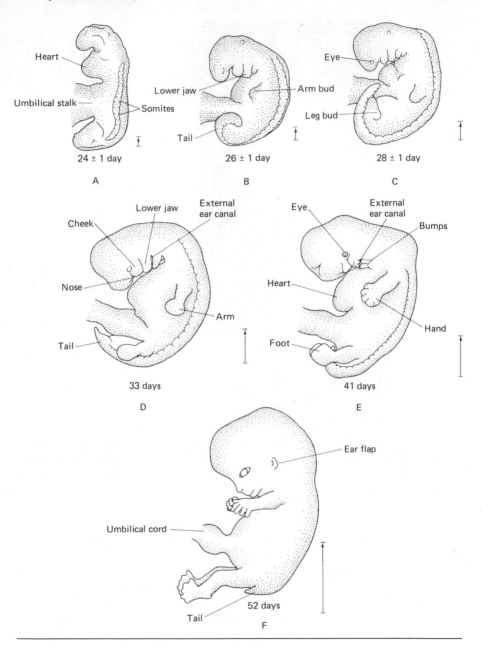

Figure 11–2 Human embryos between 3½ and 4½ weeks of age. The arrows to the right of each figure indicate the actual size. (Adapted from Moore: *The Developing Human.* W.B. Saunders Co., Philadelphia, 1977.)

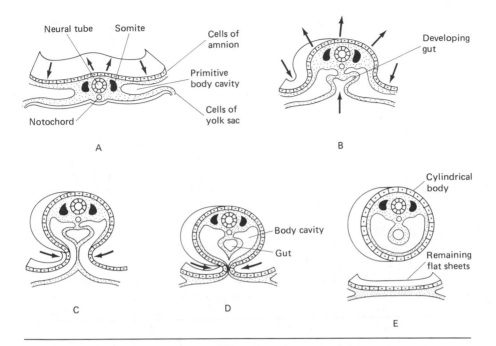

Figure 11–3 Development of a cylindrical body from the flat layers of tissues of the germinal disc. The directions of movements of the tissues are indicated by arrows. The amnion is above the embryo, and yolk sac is below.

endoderm, folds into a tube that is the primitive gut, and this gut is moved into the center of the newly formed cylindrical body (Fig. 11–3B and C). This folding process starts in the head region and works its way down to the level of the umbilicus. A comparable series of events starts at the tip of the tail and proceeds anteriorly toward the umbilicus. When this folding process is nearly finished (during the fourth week), the head and tail are cylindrical but the middle of the body is still attached to the underlying flat layers by a large **umbilical** (body) **stalk** (Figs. 11–2A and 11–3C).

The bodies of embryos in the second month have a curled-up "C" shape because at this time the brain is bent into that shape. The brain is large enough to affect the shape of the entire embryo.

Arms and legs begin as small, simple bumps, called **limb buds,** on the side of the body of the four-week embryo (Figs. 11–2B and 11–4A). Slightly later, they develop flat, paddle-shaped tips (Figs. 11–2D and 11–4C). Then a most remarkable morphogenetic process occurs. Specific groups of cells start to die and disappear. As they do this, the tiny extremity is literally sculptured from the limb buds down to the finest details. A large group of cells around the lower part of the base of the bud die and disappear, thus carving out the armpit region (Fig. 11–4C). Another group of cells on the upper surface of the middle of the arm die, helping to shape the elbow area (Fig. 11–4D). Cells die

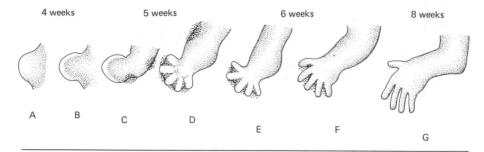

4 weeks   5 weeks   6 weeks   8 weeks

A   B   C   D
                              E   F
                                                    G

Figure 11–4 Development of the human arm. The dots represent the groups of dying cells that help to sculpture the features of the arm and hand.

in waves and disappear between the areas that will form the digits, gradually sculpturing the fingers and toes out of the flat paddles that will form the hands and feet (Figs. 11–4D to G). There are rare occasions when the sculpturing does not proceed to completion, and the individual is born with web-shaped areas of skin between the digits. This too can be corrected by minor surgery. It is an interesting paradox that programmed waves of cell death should be such an important part of shaping a living individual. The cell death process occurs in the development of the brain, spinal cord, and other structures, as well as in the arms and legs.

The extremities are beautifully formed in miniature by the end of the second month. Internal development has also progressed. Muscles, nerves, and blood vessels have been gradually created. A cartilaginous matrix for a skeleton forms and starts to convert into bone at about three months.

Development of the Face and Palate

That wonderful organ of expression, the human face, forms during the second month of development by the relatively subtle movements of a series of fleshy bars (arches) along the side of the head and neck. A pocket of skin cells, in the center of the region that will be the face and in between these bars, turns inward. This pocket becomes the mouth. The deep portion of the mouth contacts the primitive gut, and the two cavities become merged. The primitive gut, which was a tube with a blind end, now develops an opening to the outside of the body (Fig. 11–5). The rather large primitive mouth in a four-week embryo is surrounded below by a fleshy bar known as the **mandibular arch,** along the sides by wedges of tissue known as the **maxillary arches,** and above by a mass of tissue called the **frontal eminence** (Fig. 11–5A). The eyes are beginning to develop but are situated on the sides of the head. There is no nose yet, but the **nasal openings** are starting to form as two widely separated grooves at the upper corners of the mouth. It is obvious that considerable moving around of the various components of the face must occur be-

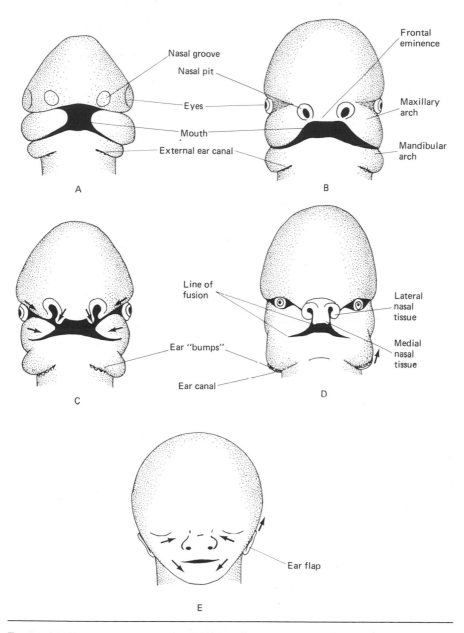

Figure 11–5 Development of the human face. *A,* Face of a 4-week embryo. *B,* The face at 4½ weeks. *C,* At 5 weeks. *D,* At 7 weeks. *E,* At 10 weeks. The directions of some tissue movements are indicated by arrows. Note that additional changes are still occurring at 10 weeks and for several months afterward.

fore they will be in their definitive location and form a face that can be recognized as human.

The various elements of the face that border the mouth start to grow toward the midline. The bulk of the frontal eminence forms the forehead. The widely separated nasal grooves deepen and become pits. The nasal openings at this time are still continuous with the mouth opening. The nasal pits, along with the raised ridges of flesh that border them, now gradually move toward the center of the face and start to project outwardly (Fig. 11–5C). The lateral borders of the nasal pits develop into the sides of the nose. The medial borders form the **nasal septum** (partition) and the central portion of the upper lip. As these changes are occurring, the maxillary arch is developing into cheeks, and the mandibular arch is forming the lower jaw.

Several important fusions between parts take place as these movements occur. The mouth gets smaller as the sides of the very wide lips fuse. The upper edges of the maxillary wedges, which form the cheeks, fuse and blend smoothly with the sides of the nose. The medial nasal tissue, which forms the center of the upper lip, and the sides of the maxillary wedges, which form the lateral portions of the upper lip, fuse together and develop into a continuous upper lip (Figs. 11–5C and D). This fusion leaves a permanent reminder: the two ridges above and to either side of the center of the upper lip. This is an important step in facial development, since if one or both of these fusions fail to occur, the resulting defect is called a **cleft lip (harelip).** This birth defect can occur on the right or the left or bilaterally.

While the components of the nose and lips are moving forward and outward, the eyes also gradually migrate to the front of the face from their initial location along the sides of the head (Figs. 11–5B to E). At the end of the second month, the fetal face is recognizably human, though many changes in proportion have yet to take place. The nose is still relatively flat at birth, and the cranium is quite large compared with the face.

The oral and nasal cavities remain continuous internally, even though the nasal pits were externally separated from the mouth when the maxillary wedges fused with the nasal processes. The internal separation occurs some time later when the **palate** develops as a consequence of another series of fusions. The roof of the mouth, the hard palate, develops from two **palatine bars,** which initially grow down into the mouth cavity (Fig. 11–6A). The tongue lies between them at this stage. Sometime during the tenth week, the tongue lowers, and the two bars rotate into the midline and fuse with each other and the nasal septum (Fig. 11–6B). Should this rather complex process not occur, the defect known as **cleft palate** results. It is one of the more common human birth defects and has been a subject of considerable experimental analysis (Chapter 17).

The external ear develops between the mandibular arch and the arch just below it. The groove that lies between the two arches deepens and forms the **external ear canal.** Meanwhile, a series of small bumps develops along the borders of this groove. Seven of these bumps grow into a whorled pattern,

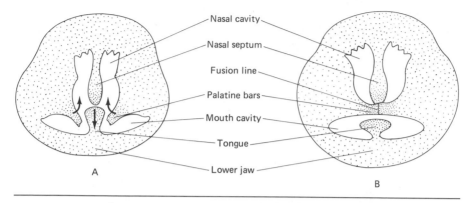

Figure 11–6 Development of the palate as shown in frontal sections, from right to left, of the heads of A, 7-week and B, 10-week-old fetuses. The direction of movement of the palatine bars and tongue is indicated by arrows.

eventually forming the external ear flap, the **pinna** (Figs. 11–2E and 11–5D and E). By the time the external ear canal forms, the middle and inner portions of the ear have developed from diverse sources and locations. These three parts—external, middle, and inner ear—now grow together and continue their development in an integrated fashion.

It is not necessarily very illuminating to say that genes control one developmental process or another, but it may be helpful to keep this in mind when you review the past few pages. A fairly large number of movements and contortions have been described in the developmental history of the face. Very subtle differences in the growth of one part relative to another can make a big difference in the final appearance of the human face. The evidence for genetic control of this process is simple; we do tend to look like our parents.

Some Internal Changes

The period between three weeks and two months is the most complex stage of our life history. During this time a primitive embryo is converted into a miniature human, about 1 inch long, as the external changes discussed previously take place. Dramatic changes are also occurring internally as new organs form literally by the hour. The new individual already has virtually every organ it will ever have by the end of the second month. Some of the changes that occur in the primitive gut illustrate how internal organs proliferate during this period (Fig. 11–7).

Numerous pockets of cells project from that simple tube, the gut. Some of these break off and form discrete organs; others remain attached. The **thyroid gland** arises as a small outpocketing of cells at the base of the tongue and then pinches off and migrates down the neck to its final location below

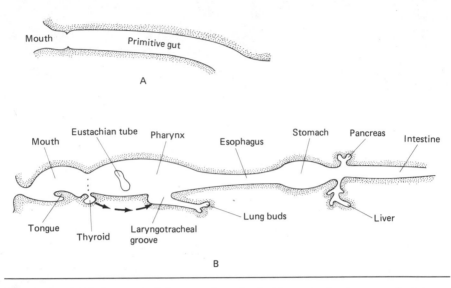

Figure 11–7 A diagram of the mouth and primitive gut of a young embryo, showing the numerous organs that form from outpocketings of this simple tube. The region where the ectoderm of the mouth meets the endoderm of the gut is shown by a dotted line. The path of migration of the thyroid gland is shown by arrows. *A*, At 3 weeks. *B*, At 5 weeks.

the **larynx** (voice box). Embryonic structures are great wanderers! If you look into a mirror and stick your tongue way out, you will see a shallow, V-shaped groove at the base with a small dimple at the point of the "V." This is where the thyroid gland originated. Off to the side of the part of the primitive gut that forms the throat are two outpockets of cells that make contact with the groove that forms the external ear. These are the **eustachian tubes,** which extend from the throat to the middle ear chamber. A little farther down on the floor of the primitive gut, a relatively large groove forms. This is the **laryngotracheal groove,** which develops into the respiratory system. The upper portion of the groove will form the larynx, and the first portion below the larynx forms the **trachea** (windpipe). Two outpocketings develop from the tip of the tracheal tube. Each will become one of the lungs by expanding into the chest and repeatedly branching (Fig. 11–8). Each generation of branches forms a progressively smaller unit of the tree of air passages of the lung, all converging onto the larynx, also the original source of the tubes. The next segment of the primitive gut below the level of the larynx develops into the esophagus. A swelling in the gut just below the esophagus becomes the stomach, and the next portion forms the intestine. Two initially simple little outpockets of cells of the gut at the border between the stomach and intestine develop into very large organs, the liver and the pancreas. Both increase in size by a process of repeated branching, similar to that of the lungs. These organs retain their connection with the gut by tubes that mark their points of origin. Both organs

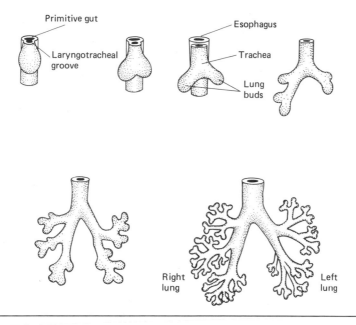

Figure 11–8 Development of the lungs resulting from repetitive branching of small outpockets, the lung buds. Many bulky organs (liver, pancreas, salivary glands, and so forth) develop in this fashion. (Adapted from Moore: *The Developing Human*. W.B. Saunders Co., Philadelphia, 1977.)

synthesize juices that aid digestion, which are secreted into the intestine via these tubes.

The foregoing discussion is just a small sample of how the simple axial organs of the primitive body at three weeks become more complex and proliferate into numerous internal organs.

The Embryo and the Fetus

The basic building units of our bodies are 160 distinctly different types of cells, each with its own structure, biochemistry, and specialized functions. These cells are arranged into tissues, and the tissues develop into organs. The **embryonic phase** is the segment of prenatal life during which the earliest development of cells and tissues occurs, the primitive body forms, and the majority of organs originate. It covers the first two months. At the end of this time, the new individual is 25 mm (1 inch) long, looks like a miniature human, and internally contains most of the organs it will ever have (Fig. 11–9G). By now you can appreciate that even the largest and most complex organs of the body have had very humble origins, such as a microscopic pocket of cells or a tiny fold of delicate tissue. At two months, most of these organs are still at a very

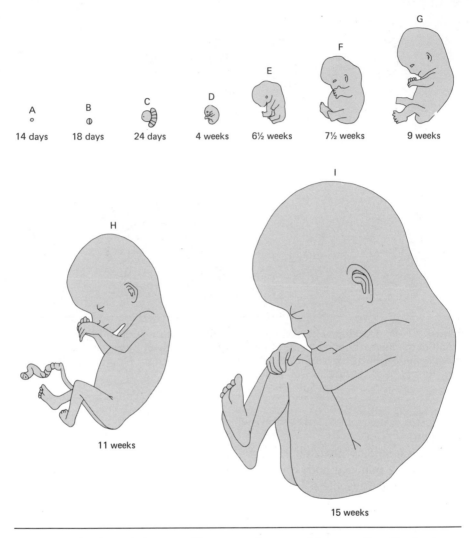

Figure 11–9 A graded series of human embryos and fetuses, all at ⅔ of natural size. (Adapted from Arey: *Developmental Anatomy*. W.B. Saunders Co., Philadelphia, 1965.)

simple stage of development, even though a few of them, such as heart and kidneys, have started to function in some fashion.

The next seven months of development are considered the **fetal phase,** and the individual is referred to as a **fetus.** In contrast to the embryonic phase, which is concerned with the origins of organs, the fetal phase is characterized by maturation and growth of the structures that are already present. These organs gradually develop from simple to complex entities and eventually become functional. Several examples of what occurs in this maturational phase are covered in the next few sections of this unit.

Growth is rapid during the fetal phase. A two-month fetus weighs about 8 grams, and a three-month fetus about 50 grams (Chapter 14). This is a sixfold increase! These weight increases are sufficiently consistent so that the age of a fetus is often estimated from its weight. A significant landmark is reached at six months, when the fetus weighs about 900 grams (2 lb). At this stage the fetus is sufficiently mature so that there is a possibility it might survive if born prematurely at this time.

Functional Development of the Nervous System

The mature central nervous system contains more than 10 billion nerve cells. Each one of these cells began as an ordinary rounded cell, which then slowly grew the extension that becomes the axon (Fig. 4–5A). The axons may grow out to as much as a meter (39 inches) in length. These billions of nerve cells in the brain and spinal cord are arranged in complex and precise patterns. Each cell develops precise interconnections with numerous other nerve cells, and each axon of a nerve cell terminates in a specific, preordained muscle, area of skin, or part of the brain or spinal cord. It is mind-boggling to realize that all of this enormously complex nervous system begins from a simple flat sheet of cells, the microscopically thin neural plate (Fig. 10–8). The details of the development of the human nervous system are surely more intriguing than any tale the human mind, itself a product of this development, could concoct.

The nervous system, in its incredible complexity, does not develop rapidly. It is no accident that the enormously complex brain is the first organ of the body to form and the last to become fully mature. The cells in most developing organs are generally at a similar level of maturity at any one time, and the entire organ becomes functional within a fairly discrete time interval. Some nerve cells of the brain and spinal cord mature and become functional at a precociously early stage, whereas others mature in waves over many years. The number of functioning nerve cells slowly and progressively increases not only before birth but also for several years afterward. Subtle maturational changes in some nerve cells can still be detected as late as age 25. This slow maturation over many years is an important feature of nervous system development because the number of mature cells and the number of functional interconnections between them are indices of the ability of the nervous system to perform complex processes.

The first nerve cells become functional at about seven weeks, when very simple reflexes can be elicited. At this time, stimulating the face will cause muscles all over the body to contract simultaneously, a phenomenon called total body response. This reflex is brought about by a small number of precociously developing nerve cells that form a transient primitive nerve network, quite unlike that of the adult.

More reflexes and progressively finer motions occur as the number of functioning nerve cells increases. The whole trunk extends at 7½ weeks, and the whole leg extends at 8½ weeks; but the knee and big toe cannot move by themselves until 10½ weeks. The fetus can swallow at three months and suck its thumb at five months. Newborns, even premature ones, have a reflex grasp that is strong enough to support their whole weight temporarily, and, even more remarkably, they can swim when placed in a pan of warm water.

Most of the reflexes just noted are controlled by the spinal cord or lower centers of the brain. The cerebral cortex, the highest part of the brain, does not start to become functional until around the time of birth. As the cerebral cortex progressively takes control over all bodily functions, as well as over other parts of the brain, many of the early reflexes are lost and must gradually be relearned. The early grasp and swimming reflexes are good examples of this phenomenon.

When does the fetus begin to feel pain during this gradual development of the nervous system? When does it experience consciousness? No one knows for sure, but most experts believe that both events occur relatively late in the fetal period.

In summary, the functional development of the nervous system takes place over a long period and is characterized by a slow, progressive increase in the capacity to perform more and more complex functions.

CHAPTER 12

Development of the Circulatory System and the Changes That Occur in It at Birth

The circulatory system is primarily responsible for delivering oxygen and food to all parts of the body, as well as being the vehicle for the elimination of soluble waste products. It is no accident that this is the first organ system to become functional, because even at very early stages the embryo needs a good supply of oxygen and nutrients.

In postnatal life, oxygen is transferred to the blood in the lungs, and food is picked up from the intestine and liver, but the fetus is not an air-breathing, free-living animal. It is a somewhat parasitic individual living in an aquatic medium and receiving food and oxygen into its blood from its mother's blood stream via the placenta. Consequently, the circulatory patterns of fetus and newborn are different. The fetal circulatory system develops in such a fashion that it can meet the needs of the fetus yet quickly change to meet the significantly different needs of the newborn at the moment of birth. These circulatory changes at birth constitute a dramatic sequence. It is necessary to discuss the development of the chambers of the heart and the major arteries associated

119

with the heart in order to understand (1) the changes that occur as the aquatic individual becomes an air-breather and (2) why developmental mistakes (birth defects) in the heart and major blood vessels are rather common. This chapter on the developmental history of the circulatory system includes a short discussion of the fetal lifeline—the umbilical cord and placenta.

The Placenta

The **placenta** serves the same functions for the embryo and fetus as the lungs, intestine, and kidneys do for children and adults. The placenta contains maternal and fetal tissue arranged in such a way that the blood streams of mother and fetus are juxtaposed but do not intermingle. Dissolved substances tend to move from regions of higher concentration into regions of lower concentration. The close proximity of blood streams permits the oxygen and nutrients, present in high concentrations in the uterine arteries, to move into the fetal blood, which has circulated through the fetus and has become depleted of these substances. Similarly, carbon dioxide and other wastes, present in high concentration in the fetal blood, move into the maternal blood stream. This interchange occurs in numerous trees of progressively smaller finger-like projections **(villi)** of fetal tissue, rich in blood vessels, which extend into pools of slowly moving maternal blood (Fig. 12–1). Any interchange of dissolved substances between two liquids is always facilitated by the presence of a large surface area between them. The breakdown of the terminal portions of the placenta into tiny villi greatly increases the surface area of juxtaposed blood streams available for interchange. This surface area is estimated to be 15 square yards in the fully developed placenta.

The umbilical cord extends from the fetus to the placenta and is about 20 inches long. Blood is delivered to the placenta from the fetus via the cord by two **umbilical arteries.** These branch off from large vessels (iliac arteries) in the fetal pelvis, traverse the wall of the abdomen to the umbilicus (navel), and enter the umbilical cord. Blood that has been freshened in the placenta returns to the fetus by a single **umbilical vein.**

The young embryo is first totally surrounded by the placental tissue. As the pregnancy progresses, the placenta becomes more localized and forms into a large disc, which weighs about 2 lb at birth. This disc usually lies across the top of the uterus. The vascular supply to the uterus increases greatly as the fetus, uterus, and placenta grow during pregnancy. The volume of maternal blood also increases nearly twofold in order to handle this extra demand (see Chapter 15).

The placenta is more than simply a place where the passive interchange of gases and nutrients occurs. The placenta synthesizes several nutrients for the fetus (glycogen, cholesterol, and fatty acids). It also serves as a very important endocrine organ, manufacturing several hormones of pregnancy (see Chapter 15).

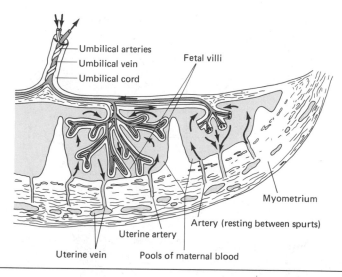

Umbilical arteries
Umbilical vein
Umbilical cord
Fetal villi
Myometrium
Artery (resting between spurts)
Uterine artery
Uterine vein
Pools of maternal blood

Figure 12–1 A diagram of the patterns of circulation of maternal and fetal blood streams in a terminal portion of the placenta. The villi are finger-like projections of fetal tissue that extend into pools of maternal blood. Note that there is no intermingling of maternal and fetal blood but that there is an extensive area of juxtaposition of blood streams, allowing for interchange of oxygen, nutrients, and waste material. The directions of the blood flow are indicated by arrows. (Adapted from Arey: *Developmental Anatomy,* W. B. Saunders Co., Philadelphia, 1965.)

Patterns of Circulation of the Adult and the Fetus

There are four chambers in the mature heart: two **atria** and two **ventricles** (Fig. 12–2). Venous blood from the entire body enters the right atrium and then the right ventricle, which then pumps blood to the lungs via the pulmonary trunk and arteries. Oxygenated blood from the lungs returns to the left atrium of the heart by way of the pulmonary veins. This constitutes the **pulmonary circuit.** The left atrium contracts and delivers the oxygenated blood into the left ventricle, which then pumps it to all parts of the body via the aorta and its branches. After the blood courses through progressively smaller arteries and ultimately through capillaries, it is picked up by various veins and then returns to the right atrium by way of the largest veins, the superior and inferior venae cavae. This constitutes the **systemic** (body) **circuit.** Both atria contract simultaneously, as do the ventricles shortly thereafter, giving the familiar "lub-dup" pattern of the heart beat. The two circuits, pulmonary and systemic, are entirely separated from each other in the adult.

The circulatory system of the embryo and fetus has three circuits, but the pulmonary circuit is not one of them. The functioning prenatal circuits are the systemic, yolk sac, and umbilical. The systemic circuit is as important and well

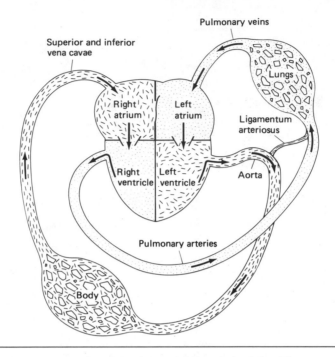

Figure 12—2 Diagram of circulation in children and adults. The systemic circuit is indicated by short dashes. The pulmonary circuit is shown by stippling.

developed in the fetus as in the newborn. The yolk sac circuit has only a transient existence in the fetus. The **umbilical circuit** (umbilical arteries, placenta, and umbilical vein) is of vital importance to the fetus. It is not needed after birth, so it is simply discarded at that time.

The lungs of a fetus are nonfunctional, so there is no need for a pulmonary circuit until birth. The basis for this circuit is established in the fetus, ready to function when the lungs start to operate. It is bypassed in the fetus by two devices that are later eliminated, one within seconds after birth and the other about an hour later.

The four-chambered heart, the arteries and veins associated with it, and the bypasses do not exist in the young embryo. These components of the elaborate system for pumping blood to all parts of the body develop gradually from a primitive set of blood vessels and a simple heart tube by a series of intricate transformations.

Development of Several Major Arteries

The arteries to be discussed are the **aorta,** which delivers blood to the body directly from the left side of the heart; the **pulmonary arteries,** which carry blood from the right side of the heart to the lungs (Figs. 12—3*B* and 12—7);

the **carotid arteries,** which deliver blood to the head and neck; and the **subclavian arteries,** which provide blood for the arms.

The major arteries arising from the mature heart originate from a series of six symmetrical pairs of arched blood vessels in the young embryo. These arches arise from the heart, extend around the sides of the neck, and merge again across the back of the neck as a pair of aortae (Fig. 12–3A). They become a single aorta at a lower level of the back. A similar set of blood vessels appears in all vertebrate embryos. In fish embryos, this set of aortic arches develops into a symmetrical set of arteries that supply blood to the gills. In human embryos (as well as those of reptiles, birds, and other mammals), an elaborate sequence of alterations occurs that eventually transforms these symmetrical arteries into the asymmetrical arteries that emerge from the heart. The changes are complex, but only a simple description of them is necessary to understand the changes in circulation that occur at birth.

The basic set of primitive arteries is diagrammed in Figure 12–3A. Note that between the six arches are two pairs of vessels that extend anteriorly into the head region. The first three arches and the aortic extensions associated

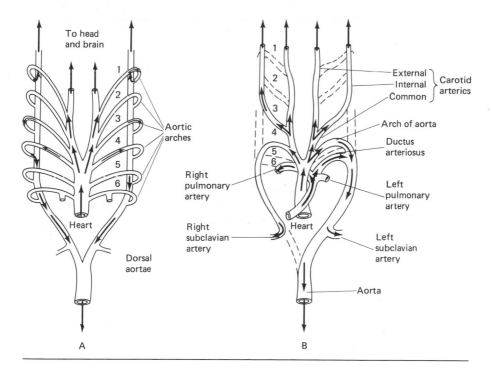

Figure 12–3 Development of the major arteries. *A,* A diagram of the aortic arch system as it might be viewed through the chest of a young human embryo, showing the basic features of the system. *B,* The major arteries of a human fetus, drawn to show how they develop from the primitive embryonic system. The segments that disappear are indicated by dotted lines. The directions of blood flow are indicated by arrows.

with them develop into the arterial system that supplies the face and brain—namely, the common, internal, and external carotid arteries (Fig. 12–3*B*). This happens because the first and second arches disappear and the segment of the dorsal aorta between the third and fourth arches on both sides disappears. The two bifurcated carotid arteries remain as the sole remnants of the first three arches. The fourth pair of arches develops in a different fashion on the right and left sides. The left fourth arch gradually becomes associated with the left side of the heart (Fig. 12–3*B*) and develops into a very large blood vessel, the arch of the aorta. It is continuous with the aorta that courses through the thorax and trunk, supplying the chest and abdomen. The fourth arch on the right becomes a small vessel (the right subclavian artery) as it loses its association with the aortic loop. The fifth arch simply disappears.

The sixth pair of arches eventually forms the arterial portion of the pulmonary circuit. This pair of arches develops a separate exit from the heart as the single primitive arterial trunk splits into two vessels. Small blood vessels arise from the middle of each sixth arch and grow downward toward the embryonic lungs. These are the pulmonary arteries. The sixth arch of the right side loses its connection with the other arches and thus carries blood from the heart directly to the right lung. On the left side, however, the sixth arch maintains a large connection with the arch of the aorta. Therefore, the left sixth arch can deliver blood directly into the aorta as well as to the left lung. This connection between the pulmonary system and the aorta is called the **ductus arteriosus.** It is a key feature of a circulatory system that can meet the needs of both the embryo and the adult. How it functions in this capacity will be discussed further on.

While these modifications of the primitive symmetrical system of arches are occurring, the heart is developing from a simple tube into a four-chambered structure. The aorta and hence the arteries that supply the body become associated with the left ventricle of the heart, and the arteries derived from the sixth arch, which supply the lungs, become associated with the right ventricle. Both systems of vessels remain interconnected via the ductus arteriosus. Meanwhile, the major veins are also undergoing an equally intricate transformation from a symmetrical set of blood vessels that return equal amounts of blood to both sides of the primitive heart to a system that returns all the blood from the body to the right atrium of the heart.

This discussion involves a fair amount of detail, but the salient features to remember are (1) that all of the major arteries associated with the heart attain their adult configuration only after some elaborate transformations and (2) that the systemic and pulmonary circuits are not separated but rather are joined together by the ductus arteriosus.

Development of the Heart

The heart of the three-week embryo is a simple, relatively straight tube (Figs. 10–8*G* and 12–4*A*). This tube then undergoes some S-shaped coiling as the

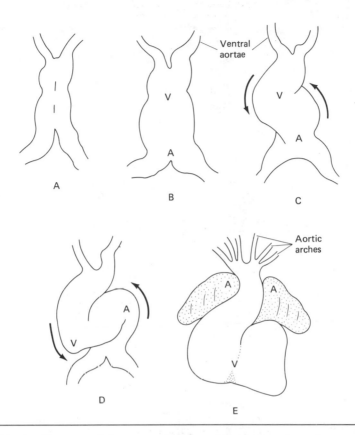

Figure 12—4 Changes in the external shape of the human heart between the third and fourth weeks of development. V, ventricle; A, atrium.

single atrium and single ventricle reverse their original orientation (Fig. 12—4). The heart has a complex external shape at four weeks but is still a single continuous tube. The divisions between atrium and ventricle are very poorly marked, and there is no separation of right and left sides (Fig. 12—5A). Compartmentalization begins as some cushiony tissue between the atrium and ventricle first narrows this region into a single channel and then develops a partition that divides the channel into right and left portions (Fig. 12—5B). A simple **septum** (partition) gradually develops over a two-week period between right and left ventricles (Fig. 12—5A to C). It consists of a stout muscular portion around the periphery and a central portion that is a simple membrane. (Failure of this membrane to form completely results in a ventricular septal defect, the most common type of congenital heart defect.)

The development of a partition between the two atria is more complex because it eventually consists of two membranes and two holes. The first to form is a thin membrane that grows between the right and left sides of the primitive atrium. This membrane does not completely separate the two atria because a hole erodes through the center (Fig. 12—5B).

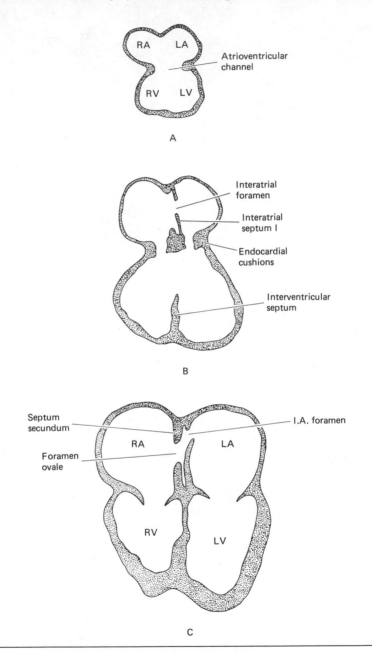

Figure 12–5 Development of the internal compartments of the heart. *A,* An early stage, four weeks. No chambers are present. *B,* Five weeks, the beginning of compartmentalization. *C,* Seven weeks—the chambers are finished.

This hole is the **interatrial foramen** (opening), which continues to allow free passage of blood between atria. A stouter second partition, the **septum secundum,** forms to the right of the first (Fig. 12–5C). It grows from the periphery to the center in the manner of an iris diaphragm and never forms a complete membrane. The hole within it has an oval shape and hence is called the **foramen ovale.** This hole is adjacent to the opening in the first septum but does not overlap it. The partition between the two atria now consists of (1) a stout and relatively inflexible second septum, (2) a flexible, membranous first septum, and (3) the nonoverlapping openings in each. This system constitutes an **interatrial flutter valve,** a very important component of the fetal circulatory pattern. This valve will open and permit blood to pass from the right atrium to the left when the blood pressure is greater in the right side of the membrane during atrial contraction. When the pressures in both atria are at least approximately equal, the thin membrane will oppose the foramen ovale and not allow passage in either direction.

The chambers of the heart are entirely formed at two months. The interventricular septum is complete. The interatrial septum, also complete, will allow blood to pass in a unidirectional manner from right to left. All of these changes are in preparation for postnatal life, since no separation into right and left chambers in either atrium or ventricle is necessary for adequate prenatal functioning of the heart. It is important to the fetus only that the blood can flow freely from right to left because the lungs are not functional and the pulmonary circuit is still considerably undeveloped compared with the rest of the body.

Circulatory Changes at Birth

The blood supply to the nonfunctioning lungs of the fetus is greatly reduced from that required by a newborn by two devices. The first of these is the ductus arteriosus (Figs. 12–3B and 12–6), which is the large interconnection between the pulmonary trunk and the aorta. Since the lungs are small and the blood vessels within them greatly compressed, blood entering the lungs encounters considerable resistance. Therefore, most of the blood that comes out of the right ventricle takes the path of least resistance, which is across the ductus arteriosus and directly into the aorta. The lungs receive enough blood to develop properly without the energy waste that would result from maintaining a complete pulmonary circuit. The draining of a substantial amount of blood from the pulmonary circulation via the ductus arteriosus sets the second bypass into motion. The fact that little blood goes to the lungs means that only a small amount of blood returns to the left atrium. The right atrium, on the other hand, still receives the much larger volume of blood that comes from the entire body. Every time the atria of the fetus contract, the pressure within the right atrium is appreciably higher than that in the left simply because it contains more blood. The interatrial flutter valve opens, and blood flows from the right atrium to the left; it then moves into the left ventricle and out into the aorta

(Fig. 12–6). Consequently, less blood flows into the right ventricle (and hence the pulmonary circulation), but the left ventricle, which pumps blood throughout the body, continues to get the exercise it needs to develop for adequate postnatal functioning.

The pulmonary circuit of the fetus is thus bypassed by these two interdependent devices. Such a bypass is adequate for prenatal life because the fetus, whose blood is oxygenated in the placenta, does not need lungs, a closed pulmonary circuit, or even a four-chambered heart. This situation changes immediately at birth, and the structure of the fetal circulatory system is such that it can change with dramatic swiftness to meet the different needs of the newborn. The structure of the lungs of an unborn infant can be compared with that of a sponge that is tightly compressed because the potential for forming a network of air spaces is present but is not expressed. The thin-walled blood vessels of the lungs are also compressed and resistant to the flow of blood. Most of the blood that flows from the pulmonary trunk, therefore,

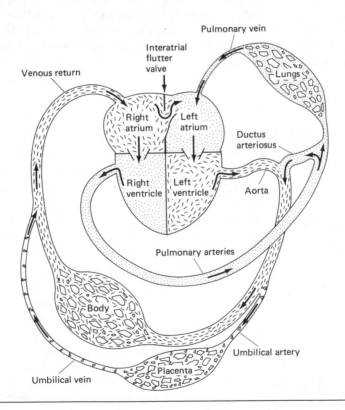

Figure 12–6 Diagram of fetal circulation. Pulmonary circulation is shown by stippling; the systemic circuit is marked by short dashes; the umbilical circuit is indicated by triangles.

passes across the ductus arteriosus and into the aorta. With the very first breath the infant takes, the tightly compressed lungs expand as they fill with air. Since the external pressure on the flexible blood vessels of the expanded lungs is also decreased, the vessels expand as well. There is less resistance to flow in the pulmonary circuit, and more blood flows into the lungs, even though the ductus arteriosus is still open (Fig. 12–7). Because there is more blood returning to the left atrium, the pressures in the two atria are approximately equal when they contract, and the interatrial flutter valve is shut off. In this manner the first breath of the newborn permanently inactivates the flutter valve, even though the tisues of the valve do not become sealed by an infiltration of connective tissue until several months after birth.

The ductus arteriosus, though still open, is no longer a very effective bypass, and it also closes off within the first hour after birth. The oxygen level within the newborn's blood stream increases rapidly within the first few minutes after birth because breathing in oxygen directly from the air is more efficient than receiving it secondhand from the maternal blood stream in the placenta. This increased oxygen level causes the muscles within the wall of the ductus arteriosus to go into a spasm, thus sealing off the ductus. It eventually becomes completely sealed by infiltration with connective tissue, but it is functionally closed shortly after birth.

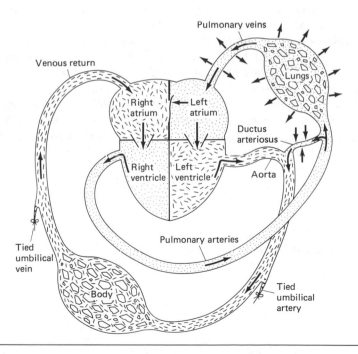

Figure 12–7 **The pattern of circulation a few minutes after birth. Compare with Figures 12–6 (fetal circulation) and 12–2 (child's and adult's circulation).**

The preceding paragraphs illustrate the way in which the fetal circulatory system rapidly converts to meet the different needs of the newborn.

What happens to the umbilical arteries and veins after birth? The external portions are, of course, cut off with the umbilical cord. The internal portions of both the umbilical arteries and the umbilical vein are large vessels that shrink after birth but persist as cords of tissue called ligaments. They can easily be seen in the adult as reminders of the prenatal phase of life. The ductus arteriosus also becomes a ligament that permanently joins the pulmonary and aortic trunks (Fig. 12–2).

Cardiac and Great Blood Vessel Abnormalities

The subject of birth defects is treated in detail in Chapter 17. However, you should realize by now that development is a very complex process, and occasionally a mistake can occur. The more involved the developmental history of an organ or organ system is, the greater is the likelihood that a mistake will occur. This generalization is particularly true with respect to the heart and the major blood vessels associated with it. The early history of the heart is one of twisting, folding, and fusion. The separation of a simple heart tube into four chambers is an intricate process. The history of the great blood vessels associated with the heart is one involving the loss of a series of segments from a primitive pattern and the formation of separations, secondary associations, and bypasses. Mistakes in the development of this system are neither common nor rare. They occur in about 1 in 150 newborns. Knowing that the development of this system is quite complex, you should not be surprised to find out that congenital defects of the heart and major arteries exhibit considerable variety. Three of the most common types are (1) an incomplete ventricular septum (p. 125), (2) an incomplete interatrial septum (usually called patent [i.e., open] foramen ovale), which occurs when the two holes of the interatrial flutter valve coincide instead of being staggered and (3) a patent ductus arteriosus, which is caused by the failure of the ductus arteriosus to close after birth. All three of these are significant defects because they prevent the complete separation of the pulmonary and systemic circuits and consequently interfere with the oxygenation of the blood. One of the great miracles of modern medicine has been the development of surgical techniques to correct many of these cardiovascular defects. Usually within every large metropolis there is a skilled staff of surgeons and ancillary help who specialize in this type of treatment.

CHAPTER 13

Development of the Reproductive System

As mentioned in Chapter 3, sex is determined at the moment of conception by a chromosomal mechanism. An egg develops into a female or male individual depending on whether the sex chromosome constitution of that newly fertilized egg is XX or XY (see Fig. 3–9). This mechanism would seem to ensure that the sex of the infant is clearly and inexorably established at conception. However, this is not necessarily so. Regardless of the genetic sex, every young embryo has all the basic structures to form a complete complement of both male and female reproductive organs. In the genetic male, the prospective female structures are normally inhibited, and the male structures are stimulated to develop. In the genetic female, the prospective male structures are normally inhibited, and the female structures are stimulated to develop fully. Sex hormones are precociously produced by the embryonic gonads and play an important role in this interesting developmental story, even to the extent of being able to reverse the direction of sexual development dictated by the genes on the sex chromosomes.

The Sexually Bipotential Stage

During its early period of sexual development, the embryo is referred to as **sexually bipotential.** This stage occurs when the human embryo is about five weeks old (½ inch long). A key structure in this early stage is the embryonic kidney because of the intricate developmental associations of the urinary and genital systems. This embryonic kidney is not the familiar kidney-shaped adult organ of the human **(metanephros)** but rather is an organ called the **mesonephros,** or middle kidney. It consists of a large number of **kidney tubules,** each of which empties into a duct, the **mesonephric duct,**

131

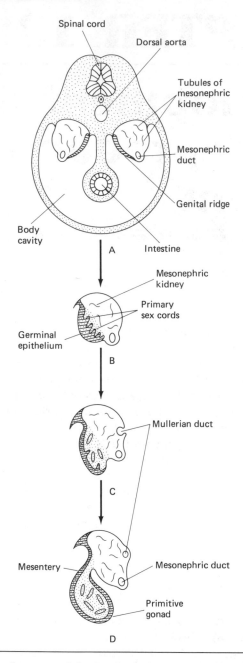

Figure 13–1 Development of the bipotential gonad. *A,* Sketch of cross section through the abdomen of a five-week human embryo showing development of the genital ridge. *B,* Initial formation of the primary sex cords. *C,* Separation of the primary sex cords from the germinal epithelium. *D,* Separation of the gonad from the mesonephros. Note that the gonad remains attached to the mesonephros by a mesentery. The müllerian duct slowly develops along the side of the mesonephros.

which then carries the urine to the bladder. This kidney is the adult kidney of fishes and amphibia. In human embryos it is well developed and functions the same as an adult kidney. The mesonephros disappears during the fourth month, after the metanephros is fully developed and functional. However, the mesonephros does not disappear completely because parts of it become converted to parts of the genital system.

The **primitive gonads** develop from the mesonephros. They arise initially as simple thickenings of the epithelium on the surface of the mesonephros close to and on either side of the midline of the body (Fig. 13–1). Cords of tissue develop from this thickened epithelium and grow inward. These are the **primary sex cords.** They eventually break loose from the epithelium from which they originated and move to the center of the growing gonad. The primordial germ cells (see Chapter 2), which have been wandering through the tissues of the embryo, now find their way into the gonad and locate in the sex cords. The primitive gonad begins to protrude from the side of the mesonephros as the primary sex cords develop, and it eventually forms a discrete structure attached to the mesonephros by a thin mesentery. The gonad at this stage is an oval organ containing primary sex cords in the middle and an epithelium on the outside. It is loosely attached to the mesonephros. This gonad is bipotential because it can develop into either a testis or an ovary.

The ducts that will eventually carry the products of the gonad are present even before the primitive gonad has formed. Two sets of ducts are present in every embryo: one set that can form the male ducts and another set that can carry the eggs from an ovary. These are (1) paired mesonephric ducts, which, in the embryo, carry the urine formed by each mesonephros down to the bladder and (2) paired **müllerian ducts,** which develop lateral to the mesonephric ducts and then swing into the midline in the pelvic region (Figs. 13–1, 13–3, and 13–4). The terminal portions of the müllerian ducts fuse in the pelvis to form a single duct. The primary reproductive organs of either sex are derived from the primitive gonad and this combination of ducts.

Differentiation of the Bipotential Gonad and Gonoducts into the Male Reproductive Organs

If the embryo has an XY constitution and conditions are favorable, the primary sex cords simply get larger and more conspicuous as they slowly become transformed into seminiferous tubules (Fig. 13–2), a process that is not finished until after puberty. The gonad is attached to the mesonephros by a strand of tissue that contains a network of fine tubules. This network, the rete testis, becomes attached on one side to the seminiferous tubules of the testis and on the other side to some of the kidney tubules within the mesonephros. Most of the tubules of the mesonephros disappear, but the few that become attached to the gonads persist and maintain their connection with the mesonephric duct. When the seminiferous tubules start to produce sperm sometime after puberty, a continuous pathway for these sperm cells is available through the rete testis,

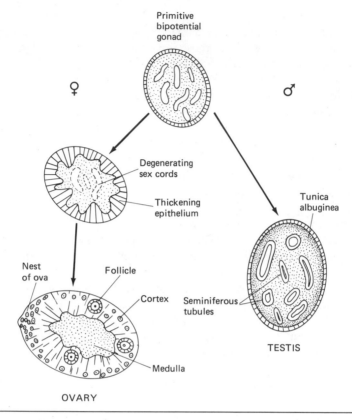

Figure 13–2 Development of the testis or ovary from the primitive bipotential gonad.

through the mesonephric tubules that persist, and through the mesonephric duct to a point below the bladder (Fig. 13–3). This change in function is associated with a change of names for the structures involved. The persistent mesonephric tubules become the tubules of the epididymis. The mesonephric duct is now called the vas deferens. The müllerian duct, which can develop into the female genital ducts, disappears in the male system except for two tiny portions. One is an insignificant little flap of tissue that remains attached to the top of the testis and is known as the **appendage of the testis** (Figs. 4–10 and 13–3). The other persistent remnant is a small tubule from the terminal portion of the müllerian duct, which is found in the middle of the prostate gland. This is called the **uterus masculinus.** Neither the appendage of the testis nor the uterus masculinus has any functional significance in the adult. They are simply reminders of the way in which the male genital ducts are derived during the bipotential stage.

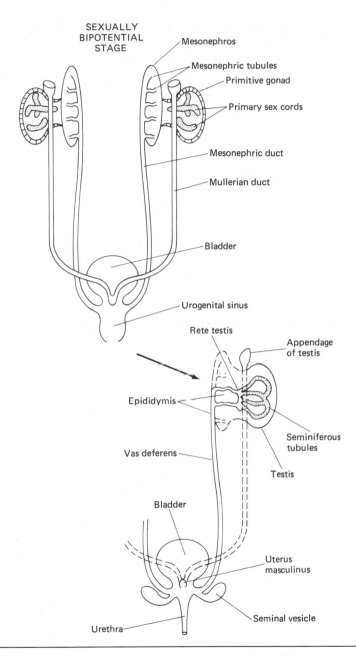

Figure 13–3 Diagram showing how the male reproductive organs develop from those structures present in the sexually bipotential embryo. Structures that disappear are indicated by dotted lines.

Differentiation of the Bipotential Gonad and Gonoducts into the Female Reproductive Organs

Chronologically, differentiation of the female structures in the bipotential embryo starts about two or three weeks later than it does for the male structures. If testes do not begin to form, the primary sex cords of the primitive gonads start to break down. The germinal epithelium, the layer of cells that originally gave rise to the primary sex cords, undergoes a renewed burst of activity and forms a very thick layer of epithelium around the gonads (Fig. 13–2). Thus the gonad that develops has a thickened outer portion, called the cortex, and an inner portion of connective tissue, called the medulla. The primordial germ cells are released from the degenerating primary sex cords, move to the cortex, and start to form egg cells.

The mesonephros and its duct disappear in the female system, leaving behind a few connective tissue strands. Here again, these strands are of no importance to the adult except as embryonic reminders of the history of this system. The müllerian ducts, which disappear in the male embryo, develop into the oviducts, uterus, and vagina in the female embryo. The portions close to the ovaries develop a large opening and finger-like folds. The next portions form the oviducts. The fused portion of the two ducts thickens and differentiates into the uterus (Fig. 13–4). A small plate of cells from the periphery grows internally to meet the uterine cervix and forms the vagina. In species of mammals that produce large litters, the uterus may develop from unfused portions of müllerian ducts, thereby giving rise to a bifid, or double, uterus. Occasionally, incomplete fusion of the embryonic müllerian ducts in the human will produce a double uterus and even, on rare occasions, a double vagina. Thus, that which is a normal situation in some mammals may, by a variation of the developmental process, occur in humans as an abnormality. This abnormality, however, is a relatively unimportant one because neither a double uterus nor a double vagina is disfiguring or a barrier to pregnancy.

The important fact to remember is that the capacity to form the structures of either sex is present in every embryo regardless of its genetic constitution. The primitive gonad and the primitive sets of gonoducts can form the reproductive organs of either sex by means of relatively slight changes. Portions of the ducts of the opposite sex remain as tiny vestigial structures. The differentiation of male and female structures during the bipotential embryonic state is summarized in Table 13–1.

Development of the External Genitalia

As with the gonads and gonoducts, there is but one set of embryonic structures in the bipotential embryo from which the external genitalia of either sex can be derived. One of these structures is the **urogenital sinus,** a space that

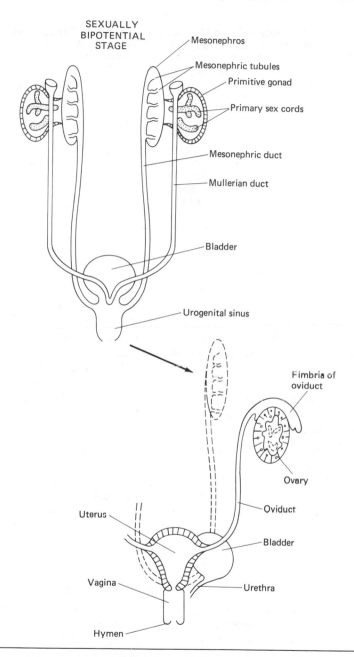

SEXUALLY
BIPOTENTIAL
STAGE

Mesonephros

Mesonephric tubules

Primitive gonad

Primary sex cords

Mesonephric duct

Mullerian duct

Bladder

Urogenital sinus

Fimbria of
oviduct

Ovary

Oviduct

Uterus

Bladder

Vagina

Urethra

Hymen

Figure 13–4 Diagram showing how the female reproductive organs develop
from those structures present in the sexually bipotential embryo. Structures that
disappear are indicated by dotted lines.

Table 13—1 Development of the Male and Female Structures Formed from the Basic Structures of the Sexually Bipotential Embryo

Structure in the Female	Structure in the Bisexual Embryo	Structure in the Male
Ovary	Primitive gonad	Testis
Disappear	Primary sex cords of gonad	Seminiferous tubules
Disappear*	Mesonephric tubules	Epididymis
Disappears*	Mesonephric duct	Vas deferens
Oviducts, uterus	Müllerian duct	Disappears*
Vestibule	Urogenital sinus	Urethra of penis
Clitoris	Genital tubercle	Penis
Labia minora	Urogenital folds	Midline raphe
Labia majora	Genital swellings	Scrotum

*Except for vestigial remnants.

has a slitlike opening to the outside of the body (Fig. 13–5A). It forms as a large embryonic bladder constricts into two portions. One part receives the ureters from the definitive metanephric kidney and retains its function as a bladder. The other part is the urogenital sinus, which maintains a connection with the bladder by a small tube, the urethra. Since this sinus was part of the bladder, it also receives (1) the mesonephric ducts, which once drained the embryonic kidney, and (2) the fused portion of the müllerian ducts. The urogenital sinus lies within an external projection, the **genital tubercle,** which is equally conspicuous in both sexes. The external margins of the urogenital sinus are marked by a pair of **urogenital folds.** Lateral to the tubercle is a pair of conspicuous **genital swellings.** In summary, the starting point for the development of the external genitalia consists of a genital tubercle, which has a slitlike opening along its base, the urogenital sinus. This opening is flanked by the urogenital folds. Internally, the sinus has four tubes entering it: the urethra, two mesonephric ducts, and the fused portion of the müllerian ducts. The tubercle is flanked by a pair of genital swellings (Fig. 13–5A).

The genital tubercle grows into a penis in a male fetus. The urogenital folds along its base fuse in a zipper-like fashion, converting the urogenital sinus, an internal space, into a closed tube. This tube now becomes the urethra of the penis (Figs. 13–5B and C). It is a continuation of the short urethra that originally extended from the bladder to the sinus. The line of fusion is marked by a persistent midline **raphe.** The base of the urethra receives the mesonephric ducts, which can now be properly called the vasa deferentia (plural of vas deferens). At this junction the ducts develop large evaginations that become the seminal vesicles. A series of about 60 small evaginations from the base of the urethra slowly develops into the prostate gland. The genital swellings to either side of the tubercle become hollow and form large pockets, the scrotal sacs, which are spaces that receive the testes as they descend from their abdominal position during the seventh month.

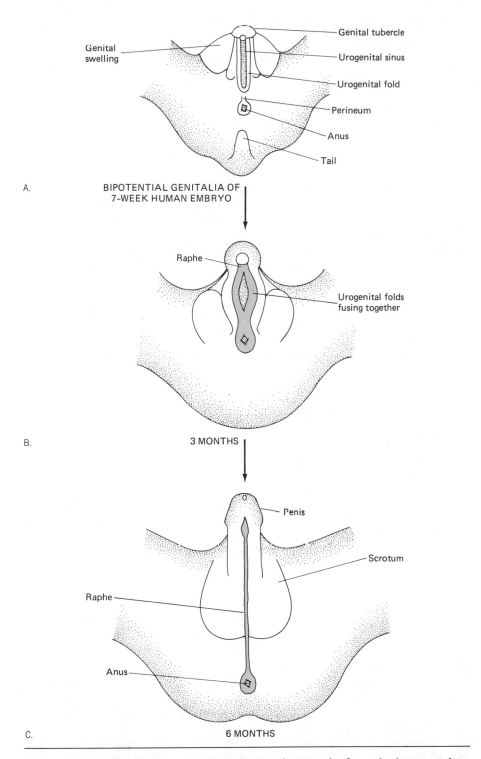

A. BIPOTENTIAL GENITALIA OF 7-WEEK HUMAN EMBRYO

Genital swelling

Genital tubercle

Urogenital sinus

Urogenital fold

Perineum

Anus

Tail

B. 3 MONTHS

Raphe

Urogenital folds fusing together

C. 6 MONTHS

Penis

Scrotum

Raphe

Anus

Figure 13–5 Development of the external male genitalia from the bipotential rudiments of the embryo.

139

The genital tubercle in the female fetus remains relatively small and develops into the clitoris, the homologue of the male penis (Fig. 13–6). The urogenital sinus, the opening that receives the urethra and the gonoducts, remains open and becomes the vestibule. It still receives the urethra as one opening and the vagina as the other. The urogenital folds, which surround the urogenital sinus, remain as small folds of tissue known as the labia minora. The genital swellings, which form pockets in the male fetus, are infiltrated with connective tissue and fatty deposits in the female fetus and become the labia majora (Fig. 13–6B and C). The development of the external genitalia is somewhat simpler in the female fetus than in the male fetus. In their final state, they more closely resemble the primitive structures of the sexually bipotential stage. The important point to remember, however, is that the male and female genitalia, despite their significant differences, have their origins in a common set of rudiments. The development of these rudiments is summarized in Table 13–1.

Causal Analysis of the Development of the Reproductive System and the Role of Hormones

The genetic sex of the individual is determined at the moment of conception, but the physical sexual features are not inevitably established at that time because the sex structures do not develop until relatively late and from a common starting point. Many different types of experiments with amphibian, chick, and mammalian embryos have repeatedly demonstrated that the development of either male or female structures does not necessarily depend upon genetic sex. A genital tubercle, under appropriate experimental conditions, can be made to develop into either a penis or a clitoris, regardless of whether the cells of that structure contain XX or XY sex chromosomes. The müllerian duct, which forms the oviducts and uterus in the female fetus and disappears in the male fetus, can be made either to persist or to disappear, irrespective of whether the cells of the duct contain XX or XY chromosomes. The primitive gonad itself can even be made to develop into either a functional testis or an ovary, regardless of whether it came from an XX or XY embryo. On what, then, does the development of the male or female reproductive system depend?

There is a balance in the genetic tendency to develop in either the male or the female direction during the sexually bipotential stage of the mammalian embryo. In an embryo of XY constitution, development first proceeds in the direction of maleness by the activation of a region on the Y chromosome known as the HY antigen locus. In some manner, still not well understood, this activation leads to the formation of a testis. This is a critical step because the testes start to secrete testosterone and some other hormones at a precocious stage. The presence of these male hormones, in turn, causes the mesonephric duct to develop into the vas deferens. A unique (nonsteroid) hormone produced only by the embryonic testis causes the cells of the müllerian duct to die. It is called the müllerian inhibitory hormone. The testosterone of the em-

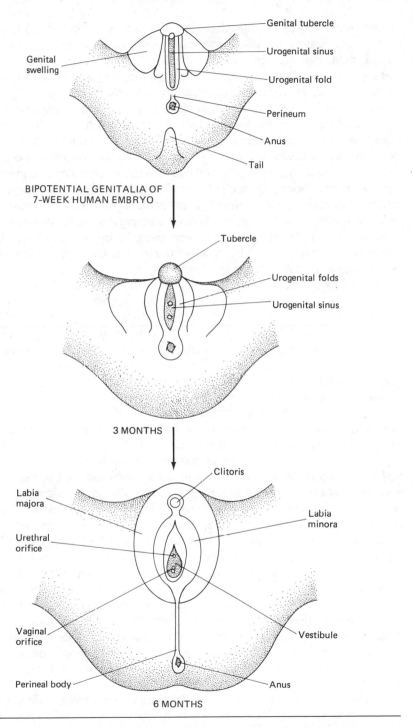

Figure 13—6 Development of the external female genitalia from the bipotential rudiments of the embryo.

bryonic testis initiates the development of a prostate gland at the base of the urethra and actively causes the genital tubercle to grow into a penis. It is apparent, then, that the genes of the XY chomosomes have their initial effect in causing the undifferentiated gonad to develop into a testis. Then the testis takes over and, by virtue of the secretions it produces, causes the gonoducts, external genitalia, and accessory glands to develop in the male direction.

Development in the female direction, in contrast, is relatively passive. If a testis does not form during the seventh week, the primitive gonad starts to develop into an ovary. It has been clearly shown that further development in the female direction occurs primarily because there are no male hormones, not because female hormones are present. The mesonephric duct disappears, the müllerian duct continues to develop into oviducts, uterus, and vagina, and the external genitalia develop in the female direction because of the absence of male hormones. These facts were demonstrated by injecting hormones into experimental embryos and by growing portions of the reproductive organs in tissue culture in the presence or absence of various sex hormones. The male hormone in mammals can induce sex transformation; female hormones cannot. This concept was elegantly confirmed by experiments in which castration was performed on rabbit and rat embryos at the sexually bipotential stage in the uterus. The development of ducts and external genitalia was always in the female direction, regardless of genetic sex. These experiments demonstrated that the presence of the testis and its secretions is the all important factor in the creation of a male individual from the sexually bipotential embryo. In the absence of a testis and its secretions, a female reproductive system develops.

The relevance of these animal studies to humans has been demonstrated in several ways. There are two rare genetic disorders that reverse the direction of the genetic sex established at fertilization. In the testicular feminization syndrome, genetic males have female external genitalia. This results from an inability of the genitalia to respond to the testosterone secreted by the embryonic and fetal testes because testosterone receptor sites on the surfaces of target cells are not present. This condition has been referred to as genetic castration. In congenital adrenal hyperplasia, an enlarged adrenal gland secretes large amounts of testosterone and causes masculinization of the genitalia of genetic females, usually an enlarged clitoris and fusion of the labia majora to form a scrotum-like organ. Some women have also been treated inadvertently with male hormones for various disorders early in pregnancy and have later given birth to genetic females whose genitalia had been partially transformed into male structures (Fig. 13–7). Such partial sex transformation results in a **pseudohermaphrodite,** an individual with a mixture of male and female genitalia. A true hermaphrodite has, by definition, both ovarian and testicular tissue. Hermaphroditism is a great rarity in mammals in both nature and the laboratory, but some human cases have been documented.

The hormonal basis for sexual development is slightly different in other groups of vertebrates. Male and female hormones are equally effective in sex transformation in frogs. The female hormone is the important one in birds.

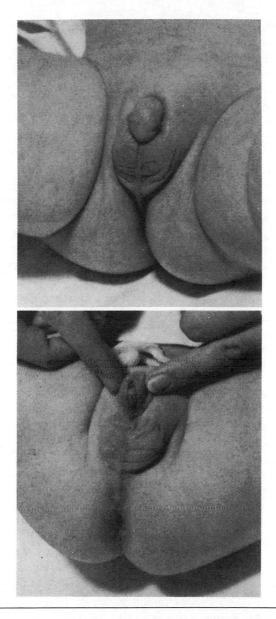

Figure 13—7 Masculinization of a genetic female infant caused by the administration of male hormones to her mother during early pregnancy. Although the genitalia seem to be those of a male, the "penis" is actually a clitoris that opens at the base into a small vestibule, and the "scrotum" does not contain any gonads. (From Moore: *The Developing Human.* W. B. Saunders Co., Philadelphia, 1977.)

Consequently, the transformation of the sexually bipotential bird embryo into a female individual is an active process, whereas development in the male direction is passive. Since the early development of mammals occurs within the body of the mother, all mammalian embryos are exposed to an environment rich in female hormones. Therefore, the evolution of a mechanism dependent on the presence or absence of the male hormone was a necessary adaptation to the intrauterine environment.

The fact that the male fetus will be secreting a significant quantity of male hormones provides a basis for determining prenatally the sex of the unborn. Toward the end of pregnancy, some subtle and temporary changes in hair pattern that reflect male tendencies can sometimes (but not always!) be noted in a mother carrying a male child. Specifically, a few pubic hairs may grow in a line from the pubis to the umbilicus. Some hairs may also grow around the areolae of the breasts. These changes are usually more noticeable in dark-haired women than in natural blondes. Sometimes these signs are searched for a little too diligently. The method is not foolproof and thus should not be taken too literally.

If you are philosophically inclined, you can debate with your friends about the significance of the fact that male development is more complex and involves more active transformations than does female development. But lest you consider this complexity an indication of natural superiority, remember that femaleness may be considered the more natural state because it is closer to that of the embryo, with maleness as an aberrant state. Whatever your choice of viewpoint, remember the basic biological fact that our sexual organs are largely what they are because of the kinds of hormones to which they were or were not exposed in prenatal life.

There is more to the totality of sex than chromosome composition or even the presence of a particular type of reproductive system. The development of sexuality and sexual orientation involves complex, prolonged, and largely psychological processes (see Chapter 19). Some recent experiments suggest that these processes, too, can be affected by prenatal and postnatal exposure to hormones. There are three times in the life history of a male individual when there are peaks of testosterone synthesis. The first of these peaks occurs during the fourth to sixth weeks after fertilization and is responsible for the changes described previously. The second is obviously at puberty. There is also a third peak, which extends from the fourth month prenatally to two months after birth. The significance of this third peak of testosterone synthesis was perplexing in the past, particularly because at this time the hormone was demonstrated to become directly attached to nerve cells in several different parts of the brain. It has now been shown in laboratory animals and humans that this perinatal testosterone production imprints the hypothalamus and pituitary to release their hormones, later at puberty, in a male pattern of continuous secretion of gonadotrophins and testosterone. A lack of testosterone at birth results in a cyclical pattern of release of hypothalamic and pituitary gonadotrophins at puberty. In other words, the inborn program of the hypothalamus is female, and unless the hypothalamus is exposed to testosterone before or after birth, the female program is expressed at puberty. The potency of these chemical programming mechanisms was dramatically demonstrated by a classic experiment in which a newborn female rat was given but a single injection of male hormone. At maturity the female (1) exhibited male copulatory behavior, (2) had normal-looking ovaries but never exhibited estrous cycles, and (3) produced estrogen at a constant level instead of in a cyclical fashion. Thus a brief exposure to perinatal testosterone permanently affected this rat's reproductive endocrine system and sexual behavior patterns. Such experimental facts are of considerable interest to both behavioral and physiologically oriented scientists. The full significance of the attachment of sex hormones to brain cells remains to be determined.

Despite the intricate interweaving of events that form this complex body of ours, we can see that development is a magnificently orderly process. This orderliness is largely because embryonic organs interact and affect each other according to an elaborate and precise time schedule. Many of these organ systems work on a cell-to-cell contact basis. The mechanism discussed in this chapter depends on diffusible substances, the hormones. These spread throughout the embryonic body and initiate and coordinate the development of a large number of structures as they shape an individual of definitive sex from a bipotential set of rudiments. Some of these target organs are located some distance from the primary control center itself, the gonad, and include the brain, hypothalamus, and pituitary gland.

CHAPTER 14

A Timetable of Human Development

This outline is a week-by-week, month-by-month compilation of the main features of human development. It serves as a chronological summary of many aspects covered in the preceding units. A few items not previously covered are included for completeness. Someday, when you or your mate is pregnant, you may want to check this outline to see what's happening to the baby at each stage.

First Week
Size: microscopic
Main features:
1. Fertilization takes place.
2. Many cells develop from one cell (cleavage).
3. The fertilized egg floats freely in the oviducts (about 3 days) and uterus (about 4 days).
4. Cells are rearranged to form a hollow ball (blastocyst) consisting of an outer layer (trophoblast) and an inner cell mass.
5. Attachment to the uterus (implantation) occurs on the seventh day after fertilization.

Second Week
Size: less than ¹⁄₁₆ inch (1.5 mm)
Main features:
1. Trophoblast invades the lining of the uterus, leading to development of the placenta.
2. A series of hollow balls forms, with the blastocyst on the outside and the amnion and yolk sac on the inside.
3. Germinal disc, between amnion and yolk sac, takes shape.

146

4. Gastrulation, the rearrangement of cells to form the three basic tissues (germ layers), occurs.
5. The primitive streak and primitive node (organizer center) form in the germinal disc.
6. Neural plate is indicated by a thickening of the ectoderm.

Third Week
Size: about ⅛ inch (3 mm)
Main features:
1. Basic body pattern develops.
2. Neural groove deepens and closes in the brain region.
3. Notochord (embryonic backbone) forms in front of the regressing node.
4. Up to 15 somites sequentially form in front of the node.
5. Intestine (primitive gut) develops as a simple tube. Toward the end of the third week, outpockets of cells from the gut mark the start of several organs: thyroid, liver, and lungs. Mouth forms and meets the foregut.
6. Heart forms from two plates that fuse and form a simple tube. Heart beat begins.
7. Eye and ear (inner) start to develop.

Fourth Week
Size: about ¼ inch (6 mm)
Main features:
1. This is a period of intense organ development.
2. Node is still regressing, laying down more of the neural tube, notochord, and somites. At the end of the fourth week, all 40 somites are present. Body is C-shaped.
3. Eyes (optic cup and lens) are well developed.
4. Atrium and ventricle differentiate from the simple heart tube. Many blood vessels are present. Circulation is good.
5. Visceral arches and aortic arches are well defined.

Fifth Week
Size: about ½ inch (10 mm) from crown to rump (CR)
Weight: 2 grams
Main features:
1. Period of organ formation continues.
2. External body is well defined but still C-shaped. Heart, kidney (mesonephros), and liver form large bulges on the surface. Limb buds start to develop. Nose forms as two small pits. Tail is prominent.
3. Face and jaws start to form from visceral arches.
4. Metanephric (final) kidney starts to develop.
5. Gonads are first visible as small ridges.
6. Condensations of mesoderm indicate the first bone formation. Some muscle masses also form.

Eighth Week
CR length: 1 inch (25 mm)
Weight: 8 grams (0.3 oz)
Main features:
1. Stage of organ formation is almost finished.
2. Body is well formed, and looks like a miniature human. Head is erect, and visceral grooves and tail are gone. Mammary glands are present. Limbs are well developed into upper and lower regions: digits are clearly present.
3. Mouth is well formed, and the palate is present as two bars (these do not fuse until the tenth week).
4. Ear flap forms from a series of seven buds surrounding the external ear canal.
5. Sexually indistinct gonad forms from the genital ridge. By the eighth week it differentiates into either a testis or an ovary. Genital tubercle forms. Accessory sex structures of both sexes are present in all embryos.
6. Four chambers of the heart are formed. Flutter valve between the atria is operational.
7. A few nerve cells and muscles are functional, and some simple neuromuscular reflexes can be elicited.

Three Months
CR length: 2½ inches (85 mm)
Weight: 50 grams (2 oz)
Main features:
1. Kidney now becomes functional and starts to secrete urine. Prostate gland forms, and the notochord is degenerating.
2. External genitalia are now clearly male or female.
3. Spontaneous activity of muscles occurs: for example, the baby starts kicking.

Four Months
CR length: 4 inches (130 mm); total length: 6 inches (150 mm)
Weight: 150 grams (5½ oz)
Main features:
1. Hair appears on the head, and ears stand out from the head.
2. Head is about one third of the total body length.
3. Fingernails and toenails start to form.

Five Months
Total length: 10 inches (240 mm)
Weight: 400 grams (14 oz)
Main features:
1. Chin, nose, and ears become more prominent.
2. A downy growth of body hair, called the lanugo, appears; it is shed before or shortly after birth.
3. Heart sounds can be heard with a stethoscope.

Six Months
Total length: 14 inches (350 mm)
Weight: 900 grams (28 oz)
Main features:
1. Eyelids start to open, and eyelashes are present. Skin is wrinkled.
2. Breathing movements may occur; the child sometimes can live if it is born at this time.

Seven Months
Total length: 16 inches (400 mm)
Weight: 1500 grams (3 lb)
Main features:
1. Fingernails and toenails are well developed.
2. Testes start to descend.

Eight Months
Total length: 18 inches (460 mm)
Weight: 2700 grams (6 lb)
Main features:
1. Subcutaneous fat is deposited, giving the fetus a plump look and smoothing out the skin.
2. Lanugo hairs are usually lost.

Nine Months
Total length: 20 inches (500 mm)
Weight: 3500 grams (8 lb)
Main features:
1. Testes are in scrotum.

CHAPTER 15

Pregnancy, Labor, and Delivery

The Physiological Changes of Pregnancy

Chapters 10 through 14 deal with the development of the embryo and the fetus. In a discussion of pregnancy, the primary emphasis is on the changes in the body of the woman who is carrying the developing baby. Pregnancy is a time of profound physiological changes in the mother and of adjustment to these changes, along with the events occurring in the uterus. Most of these alterations start during the first few months of pregnancy, initiated by the hormones of pregnancy. For example, increases in the rate and depth of breathing and increases in the amount of blood that passes through the heart may be considered adaptations of the mother's body to a rapidly growing baby during mid and late gestation. However, these physiological changes are hormone-induced and start to occur during early pregnancy, in preparation for later demands.

The Hormones of Pregnancy

It has already been demonstrated how many cyclical changes in female physiology are regulated by hormones. Pregnancy is a drastic change in the cyclical pattern, and it too is regulated to a great extent by hormones. Both estrogen and progesterone levels gradually increase during pregnancy to rather high levels. These elevated hormonal levels are necessary to maintain the fetus and placenta and to stimulate the growth of the uterus and breasts. The initial source of these hormones is the corpus luteum of the ovary. Early in preg-

150

nancy, the placenta begins to secrete significant amounts of both hormones and, by the second trimester, becomes virtually the sole source of them. The adrenal glands of the fetus aid the placenta in this function by supplying some hormone precursor molecules.

This increased hormone production is stimulated by a chemical signal from the placenta called **human chorionic gonadotrophin,** which is abbreviated **hCG.** A very important hormone, it is synthesized by the blastocyst a few days after fertilization. This hCG gets back to the ovary and is responsible for preventing the degeneration of the corpus luteum, which normally occurs in a menstrual cycle. Thus, hCG is the switch that shifts the body's endocrine system into a pregnancy status. This hormone continues to be secreted in quantity by the placenta; it (1) maintains the corpus luteum and (2) stimulates the synthesis of ovarian estrogen and progesterone until the placenta takes over this function. The blood levels of hCG become very high during the second month of pregnancy and then drop and stay at a moderate level for the duration of the pregnancy (Fig. 15–1). During this early peak hCG is particularly useful for the diagnosis of pregnancy by a simple blood or urine test (p. 153).

Another hormone formed in the placenta is **human placental lactogen** (sometimes also called human chorionic somatomammotrophin), which is abbreviated hPL. The blood level of this hormone steadily increases throughout pregnancy. This hormone, along with estrogen, which is present in abundance during pregnancy, is responsible for stimulating the growth of the breasts and preparing them for lactation. However, premature secretion of milk is prevented by the progesterone of pregnancy, which inhibits lactation.

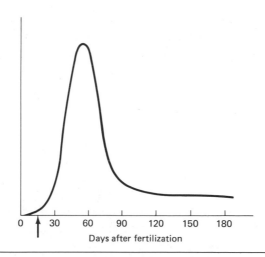

Figure 15–1 Blood levels of hCG in early pregnancy. The arrow indicates the time when an expected menstrual period is usually missed.

Skin Changes

The pituitary hormone that controls skin pigmentation is stimulated by pregnancy. It, in turn, causes a gradual darkening of the skin of the vulva and face and of the areolar areas of the breasts. Light brown patches may also develop on the forehead and cheeks. This "mask of pregnancy" is most apparent in brunettes. Most of this skin pigmentation disappears after delivery, but the vulva and areolae tend to remain darker than they were before pregnancy.

Reddish stretch marks develop across the abdomen in the last trimester in about 50 percent of women. These diminish in size and coloration after delivery but do persist.

Circulatory Changes

One of the more dramatic changes of pregnancy is the increase in blood volume of approximately 6 pints. This can be a 100 percent increase for a small woman and a 40 percent increase for a larger woman! The blood volume increases gradually throughout pregnancy. The heart prepares for this early in pregnancy by slight elevations in blood pressure and heart beat rate and by a very significant rise in cardiac output (the amount of blood that passes through the heart every minute). This great increase in blood volume does eventually put additional strain on both heart and kidneys, but normally there is a considerable reserve in the capacity of these organs.

An increase in blood volume is also, of necessity, accompanied by a rise in the number of red blood cells. Hemoglobin synthesis in red blood cells requires large amounts of iron. Sufficient quantities of this mineral are not always supplied by an ordinary diet, so iron supplements are commonly prescribed.

There is also a definite increase in tissue fluids (extracellular fluid) in all pregnancies and, with it, a tendency toward **edema.** This is a noticeable excess of extracellular fluid that causes swelling of tissues, particularly around the ankles. A dietary regimen that avoids excessive salt intake and ensures an adequate protein supply helps avoid edema.

Growth of the Uterus

The weight of the uterus increases 15-fold during pregnancy, owing primarily to growth of the myometrium. This increase in muscle volume does not result from a greater number of muscle cells but primarily from a 100-fold increase in the volume and length of existing cells. Another interesting facet of uterine growth is that the muscles of the fundus and top two thirds of the uterus develop more than those of the lower part. If the muscles of the bottom of the uterus were to contract during labor with as much vigor as those of the top, the net effect on propelling the baby would be zero. It is the muscle of the upper uterus that prepares for the major share of the work of labor.

Weight Changes of Pregnancy

Obstetricians disagree on what constitutes an "ideal" weight gain for pregnant women, but a gain of 20 to 25 lb is about average. About 9 to 12 lb of this is the weight of the baby plus placenta and membranes. Two lb are attributable to the growth of the uterus and another 1½ lb to the growth of the breasts. Toward the end of the pregnancy, the amniotic fluid can weigh 2 lb and the increased blood can weigh 6 lb.

Summary of Physiological Changes

The preceding examples are just a few of the complex changes that occur in the body of a pregnant woman. Some are profound and some are trivial. Most of them are affected by hormones that increase in both concentration and variety early in pregnancy. Even many of the changes that one might suspect would just be responses to the physiological demands of a large, growing mass of baby, such as increased breathing rate and cardiac output, actually occur early and are hormone-mediated effects that prepare the woman's body for the additional work load to come. Thus the first three months of pregnancy constitute a period of considerable alteration in a woman's body and of adjustment to it. It is not surprising that these changes are often accompanied by side effects such as nausea and profound fatigue.

The Diagnosis of Pregnancy

The obvious first sign of pregnancy is the absence of a menstrual period for ten or more days past the day it normally begins. The missed period is often accompanied by (1) frequent urination, (2) morning nausea, and/or (3) tender nipples and breasts. If, in addition to the delayed menstrual period, at least two of these symptoms are present, there is a better than 60 percent chance of pregnancy. A physician can detect additional signs of early pregnancy in the appearance of the cervix and in the size, shape, and consistency of the uterus.

The most reliable way to detect early pregnancy is to test the blood or urine for the presence of human chorionic gonadotrophin (hCG). If performed correctly, this test can be 95 percent accurate. The test for pregnancy formerly used involved the injection of a sample of urine into a mouse, rat, rabbit, or frog and then an examination of the animal's reproductive system some time later to detect the presence of hCG. Current tests are based on an immunological (chemical) reaction and take only a few minutes to 2 hours. These tests can be done on a sample of either blood or urine, since the hormone is rapidly excreted. Note in Figure 15–1 that hCG is present in the blood by the second week after fertilization but in very low concentration, is present in significant quantity 30 days after fertilization, and peaks about two weeks later. The standard office test is fairly accurate once the concentration starts to rise. This happens 30 days after fertilization, which is about 6 weeks after the last menstrual period and about 2 weeks after a menstrual period is missed. However, since the quantity of the hormone is still very low, a false-negative result can be obtained. Therefore, a negative result and a continued absence of menstruation should be followed by another test a week later, when hCG levels are at their peak. Pregnancy test kits are currently available in drugstores for home testing of urine for hCG. These are quite reliable, but, again, a false-negative result early in pregnancy can occur. Therefore, a negative test result should not be considered definitive, and the test should be repeated at least a week later. Occasionally, these tests can also give false-positive results.

The sensitivity of pregnancy tests has recently been improved by a new development, the radioimmune assay. The hCG can be reliable detected in a blood sample as early as one or two weeks after fertilization, at or just before the time of the missed menstrual period. However, this test is more expensive than the standard test and must be performed in laboratories that are specially equipped for this type of testing.

Occasionally, situations and individuals can fool the most skilled obstetrician despite all the available evidence. False-positive and false-negative diagnoses can be made. The incontrovertible proof of pregnancy is considered to be (1) detection of a fetal heart beat, (2) detection of fetal movements, and (3) x-ray or ultrasonic evidence of the presence of a fetus. These criteria can be applied during the second trimester of pregnancy.

The Course of Pregnancy

The average duration of human pregnancy is 280 days, or 40 weeks, from the first day of the last menstrual period. The time span is measured in this fashion because the last menstrual period is usually known but the actual fertilization date is rarely known and can only be estimated. About 40 percent of women begin labor within a period of 5 days before or after the expected date, 65 percent do so within 10 days, and 90 percent do so within 20 days.

The nine months of pregnancy are divided into three equal periods of three months each: the first, second, and third **trimesters.** The major characteristics of the first trimester are the profound physiological changes discussed previously in this chapter. Most women experience marked fatigue during this time and have a desire to sleep more than usual. About 50 percent of women experience nausea and some vomiting, especially in the morning. The vomiting may become so serious that it needs special attention, but only in about 2 of 1000 women. The uterus grows sufficiently so that in most women the top of it can be felt just above the rim of the pubis. The uterus also normally starts to have mild, painless contractions at irregular intervals during this period, called Braxton Hicks contractions. It is normal for them to continue throughout pregnancy.

The second trimester, by contrast, is relatively simple and comfortable.

The body has already adjusted physiologically to the state of pregnancy, and the baby has not yet reached a size that could create physical problems for the mother. The nausea and fatigue are almost always gone, and most women experience a heightened sense of well-being during this time. Fetal activity can usually be felt. The uterus grows upward into the abdomen, and the top can be felt just above the umbilicus by the end of the second trimester (Fig. 15–2B).

During the third trimester, the uterus can reach the lower edge of the rib cage. It also extends considerably forward (Fig. 15–2C). While walking, the woman changes her posture to compensate for this appreciable difference in weight distribution. The characteristic backward thrust of shoulders and the change in gait are sometimes referred to as "pride of pregnancy." Backaches are not uncommon as a consequence of the changed posture. The pressure of the baby on the bladder makes frequent urination necessary. A side effect of the hormones of pregnancy is a slackening of the muscle tension (tonus) of the stomach and intestine; this effect, in combination with uterine pressure on the abdominal organs, makes constipation a common problem. Heartburn, due to the occasional regurgitation of small amounts of stomach contents into the esophagus, may occur for the same reasons. The pressure of the near-term uterus on the veins that drain the pelvis and legs can accentuate problems with varicose veins and hemorrhoids in women who are prone to them. The baby can move about freely in the early and mid-stages of pregnancy, but in the last

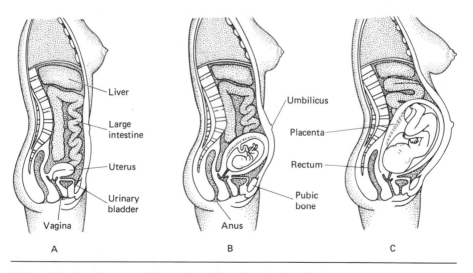

A   B   C

Figure 15–2 Drawings illustrating the relationship of the uterus to the abdominal contents during pregnancy. *A,* No pregnancy. *B,* 20 weeks of pregnancy; the uterus is up to the level of the umbilicus. *C,* 30 weeks of pregnancy; the uterus is well above the umbilicus, and the abdominal organs are considerably displaced. (From Moore: *The Developing Human.* W. B. Saunders Co., Philadelphia, 1977.)

few months it tends to assume and maintain a characteristic posture, usually one in which the head is down, the arms are crossed over the chest, and the thighs are flexed over the abdomen. The Braxton Hicks contractions become more frequent. During the last month, they serve to distend slowly the deeper portion of the cervical canal. The head descends into the pelvis during the last few weeks of pregnancy and becomes fixed into place for delivery (Fig. 15–2). This results in a dropping of the baby and uterus into the lower abdomen, an event called **lightening.** In successive pregnancies, lightening may not occur until just before the onset of labor.

Labor and Delivery

Stages

The labor and delivery process is conventionally divided into three stages: the first stage, marked by dilation of the cervix; the second stage, delivery of the infant; and the third stage, delivery of the placenta. Sometimes a fourth stage is included; this covers the first hour after delivery, a transitional period.

Onset

The stimulus that initiates labor has been very poorly understood, other than it seems to be some kind of "readiness" on the part of the fetus. Recent studies indicate that a series of interactions between the fetal adrenal and pituitary glands, along with an increased synthesis of prostaglandins (Chapter 18) by the uterus, may be the triggers.

Contractions of the uterus are necessary for all three stages of labor. Consequently, their frequency and intensity are important measures of the progress of labor. These contractions are generally accompanied by some degree of discomfort or pain (but 7 percent of women experience no pain during labor). Prelabor contractions are weak and infrequent. Labor is considered to be under way when the contractions become vigorous and rhythmical and are accompanied by some discomfort. They occur about 10 minutes apart early in the first stage of labor and about 2 minutes apart during the second stage. Each contraction itself lasts from 30 to 90 seconds with a period of relaxation between. Most of the contraction activity is in the upper two thirds of the uterus, since the primary function of the lower portion is to extend and permit the passage of the baby.

The First Stage of Labor The cervical canal is normally obliterated slowly before labor begins, but the opening of the canal is still very small and plugged with mucus. The vigorous, rhythmical contractions now push the baby's head against the opening of the cervix and progressively dilate it (Fig. 15–3A and B). It is obvious that this opening of the uterus must be fully dilated before the

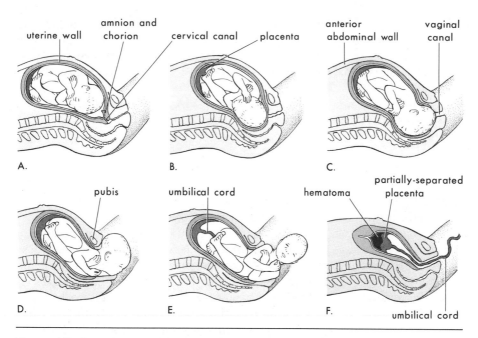

Figure 15–3 Drawings illustrating the stages of labor and delivery. *A,* An early phase of stage 1. *B,* Cervical dilation has begun. *C–E,* The second stage of labor, the passage of the baby through the cervix and vagina. *F,* The separation of the placenta from the uterus, the first phase of stage 3. (From Moore: *The Developing Human.* W. B. Saunders Co., Philadelphia, 1977.)

baby can pass through the birth canal. This process of cervical dilation takes some time. The progress of the first stage of labor is gauged by the diameter of the cervical opening, which is measured by rectal palpation or by direct vaginal examination. A blood-tinged plug of cervical mucus often escapes when the cervix is dilated by 1 to 3 cm. Referred to as a "bloody show," it is often considered a sign that labor has begun in earnest. The amnion, the "bag of waters," usually ruptures with a release of considerable fluid late in the first stage or early in the second stage of labor, but it is not unusual for this to occur just before or at the onset of labor. Voluntary contractions of abdominal muscles by the mother are of no help during this stage. The first stage is considered to be over when the cervical opening is dilated to a diameter of 10 cm, about 4 inches. The average duration is 12 hours in a first pregnancy and 7 hours in successive ones, but it may be as short as 2 to 3 hours.

The Second Stage of Labor The character of the uterine contractions changes when the cervix is fully dilated or nearly so. The contractions now begin to move the baby down the birth canal instead of merely causing the head to impinge against the cervix (Fig. 15–3C to E). Most women are aware of this change and spontaneously begin to assist the uterus by voluntarily holding their breath and increasing abdominal pressure with each contraction. Some women must be coached to do this. Either way, the **"bearing down"** process is an important part of the second stage of labor and shortens its duration.

The physician may elect to do an **episiotomy** as the baby's head begins to show and bulge against the perineum. This is an incision made from the posterior edge of the vaginal orifice and into the perineum toward the anus. This region is greatly stretched during delivery and occasionally tears. The purpose of the episiotomy is to prevent tears in the birth canal, which can be much more difficult to repair than a controlled incision (e.g., a tear that extends from the vagina to the anal canal, rare but difficult to treat). Episiotomies have been routinely performed for some years, but there is at present a more conservative attitude toward this surgical procedure. It is more likely to be considered necessary in first deliveries and less frequently so in successive ones. An alternative to episiotomy is "perineal massage," a gentle stretching of the perineum during delivery. The birth attendant can sometimes facilitate delivery by using forceps to guide the baby's head gently along the birth canal.

The largest and most difficult part of the baby to deliver is the head. The second largest is the shoulder region. Once these two parts have passed the vaginal orifice, the rest of the baby quickly comes through with considerable ease. The skull is usually compressed somewhat by the passage through the birth canal. This **molding** is normal and harmless, since the skull bones have not yet fused together and are malleable. The skull returns to its proper shape within a day or two.

Immediately after the baby is born, he or she can be placed on the mother's abdomen in a slightly tilted position to help drain excess fluid from the lungs. A gentle suctioning of the mouth and nose is normally performed for the same reason. Breathing usually starts spontaneously and quickly after birth. The lungs expand with the first breath, and full pulmonary circulation begins as the atrial flutter valve within the heart closes (Chapter 12). The umbilical cord is tied in two places about 2 inches away from the umbilicus and cut between the ligatures within a few minutes after delivery.

The second stage is over after the baby is born. It lasts for an average of 50 minutes in first deliveries and 20 minutes in successive ones.

The Third Stage of Labor   This consists of the passage of the placenta and takes about 5 to 10 minutes (Fig. 15–3F). The placenta usually becomes detached right after the baby is born as the uterus continues to contract. Further contractions help to clamp down on uterine blood vessels and minimize blood loss as well as expel the placenta.

Analgesics and Anesthetics   Some degree of discomfort is unavoidable during labor and delivery because of the nature of uterine contractions and the distention of pelvic structures. Management of this discomfort is an important aspect of obstetric care. The physician has a variety of remedies available; these range from helping to control the woman's apprehension, which can accentuate pain, to administering medicines such as tranquilizers, analgesics, and anesthetics. The analgesics that are used may be as mild as aspirin or as strong as Demerol, a narcotic. Their effect is often potentiated by the use of a mild tranquilizer or sedative. The anesthetic used may be (1) simply a local, applied to the region of the episiotomy; (2) a pudendal block, which anesthetizes the sensory nerves that supply the pelvis; (3) a spinal anesthetic, which renders the entire lower part of the body insensitive; or (4) a general anesthetic, which causes a lapse of consciousness.

These agents must be used with skill and judgment, since (1) the time period over which they are used may be considerable, (2) they can cross the placenta and interfere with the baby's initial attempts to breathe, and (3) many of them can interfere with uterine contractions and unduly slow down the progress of labor. The physician must consider, therefore, not just which drugs to use but also the timing of their administration, avoiding, for example, the temptation to use too much too soon. These agents, when used with care and good judgment, can effectively minimize pain without significantly affecting either the course of labor or the baby. The physician should also consider the mother's wishes, since many women now choose not to have an anesthetic unless absolutely necessary (p. 161).

The Mother After Delivery

Repair of the episiotomy may begin before the placenta is delivered. The vagina and cervix are examined for any lacerations. The placenta is examined after it is expelled, to make sure it is still intact and that no segment has remained in the uterus, where it could cause bleeding. For the first hour following delivery, the new mother is carefully observed for possible complications. It is important that the uterus remain firmly contracted during this period to prevent bleeding; it may be massaged occasionally to ensure this.

The first hour after delivery is the beginning of a period of readjustment in which the many physiological changes of pregnancy are now reversed. The uterus, for example, starts to return to normal size. The large amount of endometrium that developed is shed as a modified type of menstrual flow called *lochia*. This flow lasts about 10 days. The myometrium also slowly regresses, and the uterus returns to normal size by six weeks after delivery. Breast-feeding enhances these changes. The large increase of body water is also quickly reversed by a marked diuresis, which regularly occurs between the second and fifth days after delivery.

The endocrine changes are more gradual. Menstruation and ovulation usually return within six to eight weeks if the child is not nursed. The effect of breast-feeding on the pituitary gland delays their return, and most lactating women do not menstruate until three to four months after delivery. Sometimes breast-feeding can delay menstruation for up to a year or more. Such delays tend to be accompanied by a lack of ovulation and hence a decreased possibility of another conception. However, this reduction in fertility cannot be considered a reliable method of contraception.

The Newborn

In just a matter of seconds, a newborn must adapt from a parasitic, aquatic existence to that of a free-living, air-breathing individual. Most of the newborn's organs have gradually developed toward this goal. The important exception is the lungs, which must become instantly functional at birth. Breathing movements start to occur in occasional spurts sometime before birth, thus exercising the muscles used in respiration. These prenatal movements bathe the lung passages with amniotic fluid. With the first breath the newborn takes, these condensed, fluid-filled lungs take in air and expand along the entire extent of the bronchial air passages down to the microscopic terminal portions, where the vital gas exchange takes place. Proper functioning of the lungs depends in part on the sufficient maturity of the **pulmonary surfactant system,** chemicals that ensure that the newly expanded terminal portions of the respiratory tree remain open and do not collapse. This level of maturity sometimes has not been achieved in the prematurely born and can lead to **hyaline membrane disease.** Adequate pulmonary functioning also depends on circulatory changes (closure of the interatrial flutter valves and the ductus arteriosus; see Chapter 12) and activation of the respiratory centers of the brain, which control the breathing muscles.

Using a rating system called the Apgar score, the obstetrician evaluates the function of the respiratory system during the first minute after birth and then again 5 to 10 minutes later. This rating helps predict whether the baby is likely to need any aid in getting this system to function at an adequate level. The assistance may simply consist of a few minutes of oxygen administration, or it may be a more vigorous treatment. Most babies do not need any help, but the delicate nature and great importance of the beginning of respiratory activity warrant close attention by the physician.

Changes in the pulmonary and circulatory systems are dramatic, but they are not the only physiological changes that are important as the newborn adjusts to a new environment. Temperature regulation, for example, is not important to the fetus in the uterus, but shortly after birth the complex of homeostatic mechanisms that ensure this must become operational. Functioning endocrine, excretory, and nervous systems are not necessary for the fetus, but they do begin to operate gradually before birth and thus are prepared for the newborn's change in lifestyle. The intestine is even prepared for the ingestion of food by the development before birth of an array of enzymes that are ready to digest food as soon as it is presented. Thus the first few minutes, hours, and days after birth constitute a transitional period for the baby. The bases for many of these postnatal adjustments are gradually established some time before birth.

Current Trends in the Management of Labor and Delivery

There was a time when having a baby was hazardous. Even as late as 1933, the death rate for women in the United States was 620 per 100,000 births. Modern obstetric care has reduced these dangers to virtually negligible levels (about 10 to 15 per 100,000 births) by emphasizing careful prenatal care and hospital delivery, as well as by utilizing technical advances. Obstetric practice, as it evolved, included hospital labor and delivery and a nearly routine use of a general anesthetic for the final stages. Consequently, few mothers were awake to witness the birth of their children, and even fewer had the comfort of their mate's presence during labor. There is a strong trend now to counter this somewhat impersonal approach without endangering the welfare of mother or child.

The "natural childbirth" movement of today does not advocate a return to "backwoods" home delivery. It is concerned with the rational preparation of mother and father for a fuller participation in childbirth that is based on the belief that such involvement is best for the physical and emotional well-being of the child, the mother, and the family unit. Prenatal classes, which are attended by both parents, are conducted by local units of the International Childbirth Education Association (ICEA), variously known as CEA or sometimes CPEA (Childbirth and Parent Education Association), units sponsored by the American Society of Psychoprophylaxis in Obstetrics, local hospitals, and other organizations. Generally, these classes provide information about pregnancy and childbirth and train the woman to participate actively in the birth of her child. The training for this style of **prepared childbirth** consists of breathing and muscle exercises **(Lamaze Method)** that help at birth. Labor and delivery take place in the hospital, but an effort is made to have the father present as much as is possible or desired. Since it is now recognized that the routine use of a general anesthetic is neither required nor desirable in every delivery, there is usually an agreement between doctor and parents that an anesthetic will not be used unless necessary, and the mother is "awake and aware" at birth. The advantages of such a childbirth are that (1) both parents can share the joys of seeing and participating in the birth of their child; (2) the baby's respiratory system begins to function more quickly, since there is minimal or no interference from anesthetics or other drugs; (3) since mother and child are usually fully awake, the mother can cuddle and even breast-feed her child within minutes after birth. This physical contact facilitates **bonding.** Recent studies demonstrate that such contact during the first minutes and hours after a birth is a very important step in promoting the mother-child bond and has significant short- and long-term benefits for both. Opportunities for bonding should not be ignored in deliveries in which an anesthetic was given to the mother by choice or necessity.

An old profession, that of the nurse-midwife, is being revived and redefined. Nurse-midwives are specially trained nurses who stay with the mother during her entire labor and delivery. Although some perform home deliveries, most modern nurse-midwives work in conjunction with obstetricians, and the deliveries that they perform or assist with take place in a hospital setting.

Another contemporary childbirth approach is "gentle birthing." Dr. Frederick Leboyer believes that birth is a traumatic event for the baby, a drastic change from its peaceful intrauterine environment. The bright lights and noise of the delivery room add to the trauma, he believes, so the lights are dimmed during delivery, noise is held to a minimum, and the child is handled as gently as possible. An important feature is the Leboyer bath, a gentle, soothing immersion bath in body-temperature water, which is given as soon as possible after birth. Infants usually respond dramatically to this by ceasing to cry and obviously relaxing in this return to womblike conditions.

The preceding paragraphs have shown some of the ways in which the childbirth practices of recent years are being modified in a more natural, personal, and humanistic direction without compromising the safety of a hospital setting. These options are not of interest to all women. However, they are rapidly becoming more readily available for women who do want them. The agencies listed previously not only provide instruction in and facts about these methods but also serve as centers of information about local doctors and hospitals that offer these newer kinds of childbirth services.

Breast-Feeding

The breasts prepare for milk production from the earliest stages of pregnancy. The amount of glandular (i.e., secreting) tissue increases and becomes more mature under the influence of several hormones, especially estrogen, placental lactogen, and pituitary prolactin. The actual secretion of milk is prevented by the high levels of estrogen and progesterone that are present during pregnancy. The precipitous drop in these particular hormone concentrations after delivery is the signal for lactation to begin. Initially, the breasts secrete *colostrum,* a clear liquid that contains more protein and minerals but less sugar and fat than milk. Colostrum secretion is gradually converted to mature milk production over a two- to four-day period.

Both the ejection of milk from the breast and continued milk formation depend on the stimulation of the nipples by the suckling infant in an interesting series of interactions between brain and endocrine glands. Nerve impulses from the nipples pass through the spinal cord to the hypothalamus. Suckling of the nipples stimulates this part of the brain to cause the posterior pituitary to release oxytocin (p. 77). This hormone is released into the blood stream, passes to the breasts, and stimulates the smooth muscle that lines the ducts and tissues in the breasts to contract, causing the milk to be ejected from the nipples (Fig. 15–4). Continuous milk formation depends on a constant supply of another pituitary hormone, prolactin. Milk formation is also promoted by the suckling reflex, which, in this case, causes the hypothalamus to stimulate the production of prolactin by the pituitary. Conversely, a lack of nipple stimulation inhibits prolactin formation and reduces milk formation. Thus, steady nursing provides the complex stimuli for continued lactation, and ceasing to nurse will gradually end this cycle.

Breast-feeding is regaining popularity with both mothers and doctors. This is partly due to a desire on the part of many young people to return to "naturalness." But there are also some very sound medical reasons for this practice. In 1978 the American Academy of Pediatrics encouraged breast-feeding and also stated that no other nutrition was necessary for the first four to six months of the baby's life. Human milk is the most natural diet for a human baby, and its composition cannot be precisely matched by any formula. Breast milk also changes nutritionally with time to meet the changing needs of the baby. Another significant factor is that breast-fed babies are less susceptible to infections because antibodies that help fight off infections are passed along with

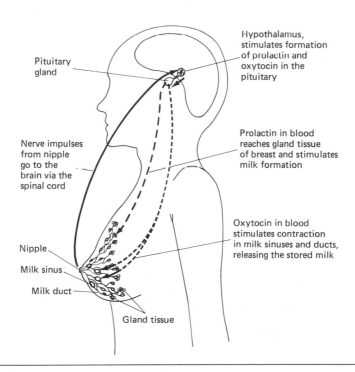

Figure 15–4 The suckling reflexes.

the milk. Suckling is also beneficial to the mother because it stimulates contractions of the uterus and helps it return to normal. Breast-feeding also facilitates bonding, with psychological benefits for both mother and child.

La Leche League International, which has chapters in most cities, and other organizations provide information and support for women who wish to try to provide their babies with this natural nourishment.

Some Complications of Pregnancy

Preeclampsia

Preeclampsia is a condition that can occur during the third trimester and is characterized by a rapid increase in fluid retention, edema, and a rise in blood pressure. If the disorder is left untreated, convulsions and other serious complications can occur. A low-salt, high-protein diet during pregnancy helps to prevent this condition. One of the aims of regular prenatal examinations is to detect the early signs of preeclampsia. The treatment for an early stage is bed rest and medication. Termination of the pregnancy is the only treatment for a woman with an advanced case. However, since preeclampsia is a problem of late pregnancy, it does not usually result in the loss of the baby.

Ectopic Pregnancy

Ectopic pregnancy results from the implantation of the blastocyst in some region other than the body of the uterus. The most common site is the oviduct, but on rare occasions the blastocyst can also attach to the ovary or intestine. The incidence is about 1 in 200 pregnancies. This is a potentially serious condition, since the embryo and its membranes are growing in a location that cannot support the growth. Eventually, the tissues rupture and abdominal hemorrhaging occurs. Ectopic pregnancies that have not reached this point are difficult to diagnose. A ruptured ectopic pregnancy is characterized by a stabbing lower abdominal pain followed by fainting. Surgical removal of the affected oviduct is the only treatment.

Spontaneous Abortion

A pregnancy that terminates spontaneously before the fifth month is called an abortion or miscarriage. They are not uncommon. About 15 percent of pregnancies abort spontaneously between two weeks and five months, and the rate prior to two weeks is estimated to be much higher. The major causes are embryonic defects (Chapter 17) and chromosomal abnormalities (Chapter 16) that are incompatible with life. A serious disease in the mother can lead to abortion. Trauma can also induce an abortion, but falls and automobile accidents are rarely responsible.

Cesarean Section

Cesarean section is a procedure by which the baby is delivered through surgical incisions made in the abdominal wall and uterus. The name was derived from an erroneous belief that Julius Caesar was born this way. This procedure is used when (1) the pelvic outlet is too small for normal delivery, (2) the uterus fails to function properly, (3) the baby's orientation would cause a very difficult delivery (e.g., a breech presentation), and (4) certain emergency situations arise. For many years, cesarean sections were performed in 3 to 4 percent of all deliveries. But recently in some areas of the United States and in some hospitals, the rate has risen to as high as 25 to 35 percent. There are some situations in which a cesarean section is mandatory, and others in which it seems merely to be a desirable option. It is in this latter type of situation that the increase seems to be occurring. This rise in the rate of cesarean sections is of concern to professional as well as community health groups.

UNIT THREE

Special Topics in Reproductive Biology

CHAPTER 16
Human Heredity

Genetics is, among other things, a science of prediction. The founder of the science of genetics, Gregor Mendel, worked with pea plants. He discovered that he could accurately predict which portion of his crop—50 percent, 10 percent, 1 percent, or whatever—would show this or that characteristic. Human genetics is also a mathematically precise science. The percentage of a large population of people who have a manifest genetic trait can be determined. The percentage of people who carry a specific hidden (recessive) characteristic can even be calculated with precision. With this information, the geneticist can predict the frequency with which a trait, hidden or visible, will occur in a future generation of a group of people. We, as individuals, are interested primarily in very small populations, namely our immediate families. Therefore, human genetic predictions are usually expressed as frequencies or odds rather than as percentages. For example, the odds that any given child of yours will be female are 1 in 2. The odds that a child may have a certain genetic characteristic are usually given as 1/4, 1/120, 1/1000, or even 1/100,000. Most sports enthusiasts understand this mode of expression very well.

Single Gene Inheritance

Some human characteristics are controlled by whole chromosomes, and some are regulated by groups of genes (multifactorial heredity). The majority of traits are controlled by single gene pairs. Each cell of the body contains 46 chromosomes as 23 matched pairs (Chapter 3). One member of each homologous pair has come from the female parent, and the other has come from the male parent. There is a point-for-point matching of genes that control a given characteristic along the length of each pair of chromosomes, though the genes at any point (locus) may not be identical. For example, assume that locus P35 on a certain chromosome is the location of the gene for dimples. George Rogers has, on the chromosome he inherited from his mother, a gene for dimples, designated *D*. At the same location on the homologous chromosome he received from his father is a comparable gene, but this one is for no dimples, designated *d* (Fig. 16–1). This is the reason for referring to chromosomes as homologous pairs rather than identical pairs. Most of the genes that control simple characteristics are in pairs similar to *D* and *d* in this example. All individuals will carry, at that locus, a combination of *DD, Dd,* or *dd* genes. Occasionally, both members of a pair are equally strong in producing an effect, but usually, one gene is more potent or is the only one that will exert the effect. This is the **dominant** gene, customarily designated by a capital letter, and the characteristic it controls is called the dominant trait. The other is the **recessive** gene, and the trait is called recessive. It is designated by a lower-case letter. In the preceding example, having dimples is dominant over the recessive trait of not having dimples. Since George has a dominant gene for dimples *(D)*, his face is dimpled. Only one dominant gene is needed for the dominant characteristic to be apparent; thus individuals who have different genes at that location *(Dd)* as well as those who have two genes for dimples *(DD)* will have dimples. Only individuals who have two recessive genes for no dimples *(dd)* will have faces without dimples. Individuals who have a pair of the same genes at a particular locus are said to be **homozygous** for that characteristic. *DD* individuals are **homozygous dominant,** and *dd* individuals are **homozygous recessive.** Those with different genes at a locus (e.g., *Dd*) are referred to as **heterozygous.**

Inherited traits are commonly noted for being able to skip generations. If a couple with dimples are both heterozygous *(Dd),* it is possible that they can each contribute a gene for no dimples to some of their children, who will then be homozygous recessive *(dd)* and have no dimples. In the parents, each of these paired genes, *D* and *d,* lies on a separate chromosome of a homologous pair. At meiosis, when chromosome reduction occurs in the formation of sperm and eggs (Chapter 3), each member of a pair goes to a different gamete. Fifty percent of all the sperm produced by the male parent will have the dominant

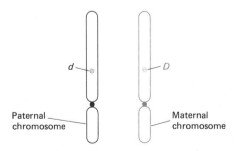

Figure 16–1 **A hypothetical pair of homologous chromosomes from an individual. Shown here at the locus marked by an asterisk is the dominant gene *D* on the maternal chromosome and the recessive gene *d* on the paternal member of the pair.**

gene, *D,* for dimples and 50 percent will have the recessive gene, *d.* Similarly, since the female parent is also heterozygous in this example, half of her own ova will have *D* and half will have *d.* It is possible, then, that at any given ovulatory period, an egg with either *D* or *d* will be released and can be fertilized with either a *D-* or a *d*-bearing sperm. The zygote that results can be (1) *DD* homozygous dominant, (2) *Dd* heterozygous, or (3) *dd* homozygous recessive. The frequency with which each of these events could occur is illustrated by the diagram in Figure 16–2. The *D-* and *d*-bearing sperm and eggs and their frequencies are indicated around the outside of the box, and the zygotes resulting from various combinations are shown within the box. A brief inspection of this diagram should convince you that there is, at any one time, an equal possibility that a *DD, Dd, dD,* or *dd* combination could occur. If enough offspring were produced, 25 percent would be homozygous dominant (dimpled), 50 percent would be heterozygous (dimpled), and 25 percent would be homozygous recessive (nondimpled). In other words, one fourth of the children of these dimpled parents would not be dimpled.

A single human family is never large enough to prove such a statistic. Consider a simple analogy to coin tossing. If you toss a coin in the air, you know that there is a 50 percent chance it will fall heads-up. If you toss it 100 times, you know that the frequency of heads-up would be close to 50 percent. You can appreciate intuitively that if you tossed the coin only five times, it is possible by sheer chance that heads-up could occur all five times, or not at all, or any number of times in between. Sperm and egg fusions, and the gene combinations that result from them, are governed by the same laws of chance. It is more convenient to express the statistical probabilities of small numbers of events, like a few coin tosses or the eye color in any one child of heterozygous parents, in terms of odds rather than percentage. Therefore, in the example concerning dimples, the odds that any particular combination—*DD, Dd, dD,* or *dd*—occurs at any one mating are 1 in 4. The possibility that the child is dimpled, either *DD* (1/4) or *Dd* (2/4), is 3 in 4. The possibility that the child is nondimpled, *dd,* is 1 in 4.

It is important to realize that if dimpled parents have a nondimpled child it does not in any way alter the possibility that the next child will be nondimpled; it is still 1 in 4. Each combination at fertilization is a separate event, just

	50% sperm *D*	50% sperm *d*
50% eggs *D*	*DD*	*Dd*
50% eggs *d*	*dD*	*dd*

Figure 16–2 **A diagram showing the possible combinations that could occur if two individuals heterozygous for dimples have children.**

as each coin toss is a separate event. A heads-up on one toss of a coin does not change the odds that the next toss will be heads-up, still 1 in 2. Having a girl at one time does not increase the chance of having a boy the next time—it is still only 1 in 2. Just as a couple may have three consecutive girls, it is possible for a dimpled couple to have several consecutive nondimpled children.

A number of basic principles of genetics have been very succinctly covered in the past few pages. Familiarity with them is required to understand the other sections of this chapter. If this is the first time that you have been introduced to this type of material, review these pages a few times before you proceed. As a check on your understanding, construct a box such as in Figure 16–2 and calculate the possible results of a mating between a person who is homozygous dimpled *(DD)* and one who is homozygous nondimpled *(dd)*. The answer is that all offspring will be heterozygous, *Dd*. Similarly, you should realize that a mating between a homozygous dimpled person, *DD*, and a heterozygote will result in offspring who are 50 percent *DD* and 50 percent *Dd*, but all of whom will have dimples.

There are many human characteristics, important and trivial, that are controlled by single genes showing dominant-recessive relationships. For example, dark brown hair is dominant over light brown, curly is dominant over straight, a clockwise crown (whorl of hair on the back of the head) is dominant over a counterclockwise crown, and hair on the middle segment of the fingers is dominant over no hair there. Dark eyes are dominant over light-colored eyes, freckles are dominant over no freckles, and a full upper lip is dominant over a thin one. The ability to roll one's tongue is a characteristic controlled by a single dominant gene. Rh$^+$ is dominant over Rh$^-$. The genes for hemophilia, Tay-Sachs disease, and phenylketonuria are recessive. Any book on human genetics will provide you with lists and tables giving facts such as these. With this kind of information, you can amuse yourself by predicting what your offspring might be like or at least by computing the odds.

Trait Frequency, Gene Frequency, and Mating Between Cousins
There is a common misconception that because a trait is dominant it may be more common than a recessive one. This is not true. The frequency with which a trait occurs in a group of people is primarily a function of the frequency with which the gene occurs. A rare gene means a rare trait, dominant or recessive. A common gene means a common trait. The frequency of simple dominant characteristics is equivalent to the gene frequency, regardless of whether the gene is common or rare. However, in the case of recessive genes with a low frequency, the possibility that two heterozygous individuals meet and mate, coupled with the possibility that only one fourth of their offspring will be homozygous recessive, tremendously reduces the frequency with which that trait appears. For example, 1 in 140 people carries a gene for albinism. This is a recessive trait, and homozygous recessives are albinos because they cannot synthesize the skin pigment melanin. A little arithmetic shows that the possibility that two individuals who are heterozygous *(Aa)* for this condition (i.e., are

carriers for this trait) should mate and have children is only 1 in 19,600 (1/140 $\times$ 1/140 = 1/19,600). In addition, we have to consider that only one fourth of their children would be albinos, *aa*, reducing the frequency to 1 in 78,400 (1/4 $\times$ 1/19,600). Thus, though the gene frequency in this case is 1/140, the frequency with which this trait occurs is 1/78,400, making this a rare trait.

All of us carry many recessive traits that are well "hidden"; that is, they are uncommon enough to be seldom apparent in families, and for the most part there is no way of knowing what they are (except for a few for which chemical tests have been devised, p. 181). We are all, therefore, carriers for many recessive traits, capable of passing them on to the next generation even though there is no sign of them in us. Most of.these are "good" or innocuous genes, but there are some that cause concern because they can produce defects, disease, or even embryonic death (recessive lethals). These hidden recessives are more likely to become expressed in the case of marriage between closely related individuals. Blood relatives obviously are likely to have many more genes in common than two people picked at random from a population. Sharing a common gene pool greatly increases the possibility that these two individuals are both heterozygous even for relatively uncommon genes and thus increases the chance for one of these hidden traits to appear. To return to the example of albinism, if two first cousins marry, and if one of their common ancestors was heterozygous for albinism (note that there are two "ifs" in this calculation), the possibility that one of their offspring is an *aa* albino increases from 1/78,400 all the way to 1/64. Modern statistical studies have shown that in marriages between first and second cousins there is a slightly greater chance that the offspring will have more problems than in marriage between distantly related or unrelated individuals. The deeply rooted sanctions against incest and marriage between cousins simply reflect the fact that such observations were made years ago without contemporary data-gathering techniques.

Other factors also influence trait frequency, so that traits do not always appear in a person who has the genetic composition for those features. The influence of some genes does not become apparent until relatively late in life, so that young people who carry the gene would not be obviously affected. For example, Huntington's disease, caused by a dominant gene, is a fatal neurological disorder that strikes in middle age; potential victims cannot be detected before then. Some genes can be modified or prevented from producing an effect because of other neighboring genes (modifiers). A person in that case would not show the trait but would still be able to pass it on to offspring. Sometimes people who have the genetic makeup for a trait do not show it for reasons that we do not fully understand (a phenomenon called lack of penetrance). Such considerations affect genetic predictions.

Sex is also a factor that can limit or influence the expression of a trait. Genes for a double vagina or large breasts can be carried by a man but obviously will never be shown by him (sex-limited traits). The prime example of a **sex-influenced trait** is pattern baldness. The trait is sometimes described as being dominant in men but recessive in women. The truth is that the trait requires the male hormone testosterone to become evident in heterozygous individuals. Heterozygous *(Bb)* women, young boys, and eunuchs (castrated men) will not lose their hair even though they have the right genetic constitution. Heterozygous men start to lose their hair slowly some years after puberty when adult levels of testosterone are reached. Heterozygous women and eunuchs start to lose their hair if they are treated with male hormones. This combination of the correct gene for a trait plus an additional factor (in this case, testosterone) shows how the expression of genes can be affected by environmental and physiological factors.

Special Genetics of the X Chromosome　The expression of genes on the X chromosome having dominant-recessive interactions is different in men and women, since women carry two X chromosomes and men only one. Such genes are called **sex-linked,** even though they normally have nothing to do with any sexual characteristic. Hemophilia, a serious deficiency of the blood-clotting mechanism that results in excessive bleeding from even small injuries, is a good example. It can be caused by several different genes, but the most famous type is one that is carried on the X chromosomes. The notoriety of its transmission results partly from the fact that Queen Victoria was a carrier who passed it on to some of the royalty of Europe and, quite literally, affected the course of history. This form of hemophilia is a recessive trait. A woman can be *HH* (homozygous nonhemophilic), *Hh* (a heterozygous carrier), or *hh* (homozygous hemophilic). Men have only one X chromosome; therefore, the single gene that they have at this locus on the X chromosome is expressed regardless of whether it is dominant or recessive. Men are either normal, if they have the dominant gene, *H*—, or hemophilic, if they have the recessive gene, *h*—. Since a man receives his single X chromosome from his mother (Chapter 3), this trait is always passed from mother to son if the woman is carrying a gene for hemophilia, that is, if the gene pair is *Hh* or *hh*. However, *hh* women are exceedingly rare because this condition almost always causes death before reproductive maturity. The male hemophiliac, therefore, has received the gene for this disease from a mother who is a heterozygous carrier (Hh) without any symptoms and who thus is usually unaware of this fact. (There are some new testing procedures that can detect this condition, however. See p. 181.) There can never be a father-to-son inheritance simply because a father does not pass on an X chromosome to his sons. A box diagram for the inheritance of this trait identifies the sex of the offspring, since it also documents the passage of the X chromosome. Such a box shows that in a mating of a normal man and a carrier woman, half of the male offspring will be hemophiliacs and half will be normal (Fig. 16–3). Note also that half of the female offspring will be carriers without symptoms.

　　　Red-green color blindness is also inherited in this fashion, as well as a number of other traits.

The ABO Blood Groups　Genes control not only visible bodily characteristics but also various aspects of body chemistry. The **blood groups,** which result from numerous chemicals circulating in the blood, are in this latter category. The chemicals responsible for the inherited blood types are called antigens, substances that can react with an appropriate antibody. The inheritance of the common blood types—A, AB, B, and O—is a variation on the simple dominant-recessive combination in that (1) three different genes, A, B, and O, determine blood type, but each person has only two of these genes because there is room for only two, one on each of the pair of chromosomes that carry this trait, and (2) the A and B genes are codominant (equally potent), and both are dominant over O, which is the recessive. These three genes control blood type in the following manner. If a person receives an A gene from one parent and either an A or O from the other, he or she is either AA or AO and is of blood type A. Similarly, a person who is BB or BO is of type B. However, an individual who inherits an A gene from one parent and a B gene from the other has both A and B antigens in the blood and is of blood type AB. The person who receives the recessive gene O from both parents is of blood type O. Thus three genes can produce four different effects.

Rh Incompatibility Everyone carries many different kinds of inherited groups of blood antigens in addition to the ABO set. **Rh negative** (Rh⁻) and **Rh positive** (Rh⁺) are another of these. An Rh incompatibility may sometimes arise between mother and fetus that results in Rh disease of newborns, also called **erythroblastosis fetalis.** Modern obstetrical practices have practically eliminated this as a significant problem, but it is still a matter of concern to prospective parents.

The Rh⁺ gene is dominant over Rh⁻; therefore, individuals who are homozygous positive, Rh⁺Rh⁺, as well as heterozygous, Rh⁺Rh⁻, have Rh-positive blood. Fifteen percent of Caucasians are homozygous recessives, Rh⁻Rh⁻, and have Rh-negative blood. If a woman who is Rh-negative has children with a man who is Rh-positive, some or all of the children may be Rh-positive. All will be Rh⁺Rh⁻ if the father is homozygous positive; 50 percent may be Rh⁺Rh⁻ and 50 percent Rh⁻Rh⁻ if the father is heterozygous. It is this combination of Rh-negative mother and Rh-positive child that can sometimes lead to incompatibility. Calculations that take into consideration the gene frequency and other chance factors such as matings of Rh-negative

	50% sperm H (X)	50% sperm − (Y)
50% eggs H (X)	HH (XX,♀)	H− (XY,♂)
50% eggs h (X)	Hh (XX,♀)	h− (XY,♂)

Figure 16–3 Box diagram showing the inheritance of a sex-linked characteristic, hemophilia, when a carrier woman has children with a normal man.

women with Rh-positive fathers indicate that an Rh-negative woman carries an Rh-positive child in about 9 percent of all pregnancies. However, the incidence of Rh disease is much lower than this (about 1/200) because other circumstances affect this type of complication of pregnancy.

Erythroblastosis fetalis is the consequence of an immune disease between an Rh-negative mother and her Rh-positive child. The disease, in order to develop, requires some fetal blood to enter the maternal blood stream. This is a relatively uncommon event during pregnancy but can occur at delivery. The Rh-positive blood from the fetus stimulates antibodies against Rh-positive blood cells to form slowly in the maternal blood stream if such a leakage does occur. The antibodies do no harm to the Rh-negative mother, but when they return to the fetal blood stream, they slowly cause the destruction of the baby's blood cells. The severity of the symptoms, which range from those that are insignificant or slight to severe jaundice and even death, depends on the concentration of maternal antibodies present and the amount of fetal blood destroyed. The "ifs" and the "slowlys" in the preceding sequence of events all tip the scales in favor of the baby. Fetal blood transfer never occurs in some women, and many Rh-negative mothers have had large families or Rh-positive children without any degree of Rh problems. The buildup of antibodies in the maternal blood stream is slow and usually requires more than one pregnancy to reach a significant level. Therefore, the first child in any incompatibility situation is almost always free of symptoms. It is only with successive pregnancies that the degree of risk progressively increases, and this can be carefully monitored. Finally, the disease develops slowly and does not become significant until late in pregnancy. Prompt treatment of the newborn in suspected cases, mostly a matter of blood transfusion, is enough to treat such cases successfully. A transfusion can even be performed before birth when a severe case of erythroblastosis is suspected. There is also a powerful new medical tool in the fight against Rh disease. A serum (RhoGAM) has been developed that, when given to Rh-negative mothers after delivery, helps to prevent the buildup of maternal antibodies.

The odds against the development of Rh diseases are strongly in favor of the baby because of the nature of the required sequence of events. Modern obstetrical practices have made this disease a virtually negligible problem.

Blood Group Genetics and Paternity Determinations   There are many types of blood groups, and laboratory tests for them are accurate and simple. These tests are widely used, not only for transfusions but also for other types of medicolegal identifications. The inheritance patterns of these blood groups are well known, and for this reason they are used in paternity lawsuits. Usually, the identity of the mother is known, and only that of the father is sought. Suppose, for example, that a mother of blood type A Rh$^-$ has a child of blood type AB Rh$^+$. The mother is of type A (genetically AA or AO); therefore, the child must have received the gene for type B from the father. He would have to have been of either type B or type AB. If the alleged father is of blood type

A or O, it can be said conclusively that he is not the father. If the alleged father is of type B or AB, it can only be said that he could be the father, not that he definitely is, since there are many B and AB men in the world. If the results of the ABO test are inconclusive, the testing proceeds to the Rh series. Since the mother is Rh^-, the child is heterozygous and had to receive the Rh^+ gene from the father. If the alleged father is Rh^-, he can be eliminated from consideration. If he is Rh^+, this again shows only that it is possible that he could be the father.

Blood testing of this kind could formerly only eliminate some suspected persons and implicate others but could not conclusively prove paternity. However, because of recent refinements using many genetically determined chemical markers in the blood cells and serum, this type of testing now has a 99 percent accuracy rate in determining if a man is the father of a child.

Multifactorial Inheritance

All of the examples in the preceding section deal with discrete traits of the all-or-none variety controlled by single gene pairs. There are, however, numerous human characteristics that are controlled by several gene pairs in combination and exhibit continuous variation. Examples are intelligence, stature, body build, skin pigmentation, longevity, blood pressure, and even the total number of fingerprint ridges. These characteristics can usually be measured, so they are sometimes referred to as quantitative traits. When a plot is made of the measurement on the horizontal axis against the frequency of occurrence on the vertical axis, a bell-shaped curve is usually obtained (Fig. 16–4). The majority of individuals are clustered near the center of the range, with progressively fewer individuals found in either direction from the center. The continuous nature of this type of variation is due to the many possible combinations that can occur in these multifactorial situations, regardless of whether or not any of them are dominant, recessive, or codominant.

With characteristics that take a long time to develop, such as height and intelligence, the effects of environment on the expression of these genes can become significant. The potential, a ceiling or a floor, for such traits is what is inherited. This potential may be nurtured to its optimum expression, or it may never be reached because of external causes. All the genes for tall height can be negated by extended childhood diseases or malnutrition. Genes for longevity becomes worthless if a person steps in front of a rapidly moving car. The effects of a combination of genes for high intelligence can also be nullified by a serious degree of oxygen deficiency during the birth process, which causes brain damage and results in mental retardation. Heredity and environment are both important in the development of many human characteristics.

The total genetic makeup of an individual is also responsible for the many variations we note in bodily responses to various environmental stimuli, such as resistance and sensitivity to diseases, drugs, and birth defects (p. 190).

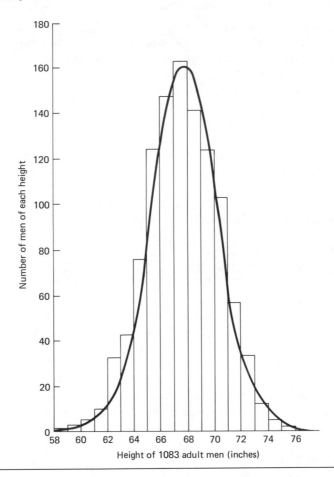

Figure 16–4 An example of a "normal curve," or curve of normal distribution, showing the heights of 1083 adult white men. The blocks indicate the actual number of men whose heights were within the unit range; for example, there were 163 men between 67 and 68 inches in height. The smooth curve is a normal curve based on the mean and standard deviation of the data. (Adapted from Villee: *Biology*, 6th ed. W. B. Saunders Co., Philadelphia, 1972.)

Chromosomal Abnormalities

Chromosomes, their significance as carriers of heredity units and their behavior during cell division, are discussed in Chapter 3. This chapter also shows how the X and Y chromosomes are responsible for controlling one important human characteristic—the sex of the individual. Because of the rather elaborate process of duplication and separation of duplicated pairs, which occurs at each cell division (mitosis, Fig. 3–4), and the extra complexities of meiosis (Fig. 3–5), it is not surprising that occasional mistakes occur. These chromosomes are of tremendous importance to the cell and the individual. Mistakes that result in chromosomal abnormalities are not well tolerated. It is not of serious consequence when a mistake occurs in a dividing skin cell or stem blood cell and the abnormality drastically interferes with the functioning of this cell, because there are millions of these cells. However, if the chromosomal abnormality occurs in the formation of a sperm or egg cell, this mistake can be passed on to every cell of the new individual and seriously affect both its development and its function. The usual fate of such an individual is an early death. Studies of human embryos that were spontaneously aborted before the third month of gestation showed that about 40 percent of them had chromosomal abnormalities that appeared to be responsible for their deaths. Chromosomal abnormalities are found in only 1 in 500 live births.

One category of these involves missing chromosomes. They are exceedingly rare because too many missing genes are incompatible with life. The lone exception to this is the missing Y chromosome, presumably because it contains relatively few genes. A portion of a chromosome can also get lost (deletions), and sometimes portions of chromosomes get mixed up (translocations). The deletions and translocations vary considerably in frequency of occurrence and in severity of consequences.

Another type of chromosomal abnormality is the presence of an extra chromosome so that a matched set would consist of three of a kind, a **trisomy,** instead of the usual pair. The most common of these is a trisomy of chromosome Number 21, which results in **Down's syndrome** (formerly called mongolian idiocy). The main symptom is severe mental retardation, but some minor physical abnormalities also occur in internal organs, eyelids, and fingerprints. Down's syndrome is by far the most common chromosomal aberration that is not lethal, occurring in about 1/700 births. This statistic alone does not tell the whole story because the incidence of trisomy 21 increases markedly with advancing maternal age. The risk in mothers younger than 30 is only 1/1500, but in mothers older than 45, the risk greatly increases to 1/30 or more (Table 16–1). Presumably, this is because ova are many years old

Table 16–1
Relationship
Between Maternal
Age and Occurrence
of Down's Syndrome

Maternal Age (years)	Risk
20	1/1900
25	1/1200
30	1/900
35	1/350
40	1/110
45	1/30
50	1/10

before they are released, and the older they are, the greater the chance that a chromosomal anomaly will develop.

The sex chromosomes are also subject to abnormalities that are not lethal. An XXY combination (Klinefelter's syndrome) occurs in about 1/1200 births, an XO (Turner's syndrome) in about 1/7000 births, and an XYY in about 1/4000. The symptoms vary from fairly severe in XO women to innocuous in most XYY men.

Genetic Counseling

Genetic counseling as a folk practice has existed for centuries. My own childhood neighborhood was quite a cultural melting pot. I remember my Armenian friends telling me that when a young couple was seriously contemplating marriage, their older relatives would meet to discuss familial defects and diseases. If the families had too many in common, the marriage would be discouraged. Geneticists have been practicing counseling on a more formal basis for more than 50 years. The counselors have available to them long tables giving statistics for various genetic problems that also take into consideration such factors as racial background, family histories, and other pertinent variables. Advice of this kind is sought for various reasons. The most common group to inquire are parents who have had one defective child and want to know the chances that a subsequent child may be affected. Human genetics is a game of odds, and the answer has always been given in terms of probabilities, whether the question arises from intellectual curiosity or is backed up by a genetic tragedy. Recently developed, powerful tools such as amniocentesis and tests for heterozygous carriers now often allow counselors to reinforce simple statistics with more specific and positive advice.

Amniocentesis is the withdrawal of amniotic fluid by puncturing the abdominal wall and uterus with a long hypodermic needle. A number of chemical tests for genetic diseases can be performed on this amniotic fluid. This fluid always contains a few cells cast off from the skin and membranes of the fetus that can be used for chromosomal study and for genetic sex determination. The tests are not suitable for mass screening of all pregnant women. However, they can be very effective when a genetic problem seems highly probable. Here are several examples.

Down's syndrome is an expensive tragedy. The mental retardation is often sufficiently severe that the child requires institutionalization. The cost in emotional anguish is also high. This problem can be diagnosed before birth (during the fourth to fifth month of pregnancy) by examination of the chromosomes of cells obtained by amniocentesis. An abortion can be suggested when there is a positive result. When is such a test warranted? It is not necessarily mandatory in families in which Down's syndrome has occurred because each case of trisomy is a new event and not normally passed from generation to generation. Nor is it particularly necessary in a pregnant woman who is 25, for whom the risk of having a child with Down's syndrome is only 1/1200. But it is highly recommended for women older than 40 because the risk is 1/100 (Table 16–1), and in women older than 45 the test should be routine. State governments have a financial interest in Down's syndrome because of the cost of the institutional care of these children. Some states provide free tests for Down's syndrome to all pregnant women older than 35. However, knowing before birth that your child has Down's syndrome is of little value to you if having an abortion is against your personal credo.

Tay-Sachs disease is a genetic disease in which a substance necessary for the development of the nervous system, the enzyme hexosaminidase A, is missing. The result is that the baby's nervous system starts to degenerate slowly at about six months. The disease inevitably terminates in death, which occurs between three and five years of age, after a prolonged period when the child is a virtual vegetable. This disease is caused by a recessive gene, *t*. The child must be homozygous *(tt)* for this condition to occur and would have to have received a *t* gene from both parents. Conversely, both parents would have to be heterozygous *(Tt)* carriers (Fig. 16–2). This disease is largely (90 percent) but not exclusively confined to Jews whose roots can be traced to southern Lithuania and northeastern Poland. A Jewish couple with no family history of Tay-Sachs disease has a 1/900 chance of having a child with the disease, but if there is a family history the odds increase to 1/50. If a couple has already had a child with Tay-Sachs disease, the parents are obviously heterozygous for this condition, and the odds that another child of theirs would have it are 1/4.

Tay-Sachs disease can be detected prenatally by testing a sample of amniotic fluid. A positive result would indicate that an abortion might be desirable. There is also a new diagnostic tool available that can detect potential carriers of Tay-Sachs disease, that is, *Tt* heterozygotes. Since this gene is most likely to be present in Jewish populations, synagogues often organize screening programs. Once an individual is diagnosed as a carrier, the hazards of having children with another carrier must be carefully considered.

There are similar tests available for heterozygous carriers of phenylketonuria, cystic fibrosis, and some forms of hemophilia. A woman who comes from a family with a history of sex-linked hemophilia could find out if she is a heterozygous carrier. If this is so, she will know that none of her daughters will be bleeders but that half of her sons will be seriously affected (Fig. 16–3). A prenatal sex determination can be done by amniocentesis. If her child is male, she could elect to have an abortion and thus avoid the possibility of having a seriously ill son with a very limited life expectancy. Although this is not a cure for the disease, it can be of help to some people.

It is important to realize that despite the emphasis on disease in the latter part of this chapter, genetic diseases are relatively uncommon and most of them are, in fact, rare. Furthermore, this is a field in which much progress is rapidly being made. Should you ever have reason to think you might have a genetic problem, your local March of Dimes office can give you the location of the nearest genetic counseling service. This society supports research and treatment centers for all birth defects, genetic and nongenetic.

CHAPTER 17

Birth Defects

Before we continue, it is very important for you to realize that almost all pregnancies produce perfectly normal babies. This chapter concentrates on the relatively few exceptional babies who show morphological abnormalities, who are mentally retarded, or who die just before or shortly after birth because of a defect in development.

Most development of new structures takes place in the short time between the third week after fertilization and the second month. This is, unquestionably, the most fragile and delicate period of existence. New organs are formed, literally by the minute, and achieve their adult relationships to each other by a series of complex interactions that must proceed according to an intricate timetable. Recognizing the complexity of embryonic development, particularly during this six-week period, one cannot help but wonder why developmental mistakes do not occur more often.

Frequency of Developmental Defects

How serious a medical problem is defective development? Relatively good statistics have been kept on human babies for some time now. Polydactyly, the presence of extra fingers or toes on the hands or feet, respectively, occurs in about 3/10,000 births. Cleft palate, a considerably more serious problem, has a frequency of about 1/1000 births, and Down's syndrome occurs in about 1/700 births. Such statistics can also be expressed in terms of defects that affect a single organ system: for example, congenital heart defects occur in about 1/200 births, and urogenital defects are present in about 1/300 births.

A more important way of phrasing the question of frequency is, What are the chances that any particular child will have a significant defect? This is a difficult question to answer because (1) some defects such as an extra or missing kidney may not be detected without diagnostic examinations that are not ordinarily done and (2) some birth defects have many gradations, and statisticians differ in interpreting the significance of some of the more minor ones (e.g., a small birthmark in the middle of the back is obviously trivial, but a port wine–colored birthmark that covers part of the face is not insignificant). Adding to the problem of evaluating overall risks is the recognition that **growth retardation, mental retardation, functional deficits** of the endocrine and other organs, and even some rare forms of cancer can have prenatal origins, and these may not become evident until some years after birth. Consequently, statistics on the incidence of human birth defects as a whole vary a great deal, depending on what is and what is not included and how thoroughly the diagnostic work is done. The figures range from 1/100 births all the way to 1/50 or even 1/25. Even the lower figures include some small abnormalities that can be ignored or easily corrected.

Causes of Birth Defects

A Historical Survey

The question "Why was my child born defective?" is a cry that has been heard through the ages. This is an emotionally compelling question that demands an explanation. Even today, if a doctor cannot supply the parents with a satisfactory answer, which is possible only about 20 percent of the time, there is a tendency to return to the old superstitions.

Prehistoric humans asked this question and left behind cave drawings and statues of deformed individuals, evidence that they considered this a matter of religious significance. The ancient Babylonians asked this question more than 5000 years ago and left catalogued descriptions of many defects in humans and animals for the purpose of divination, the foretelling of the future. Needless to say, the birth of an obviously abnormal individual was usually considered a portent of some type of bad luck or evil. The Greeks, on the other hand, thought that the gods produced defectives to amuse themselves. These defectives, special creations of the gods, were treated with enlightenment and respect. This type of tolerance was not exhibited throughout most of recorded history. Defective children, and often their parents as well, were treated with fear and suspicion. They were often ostracized and sometimes even executed.

One explanation for defects, frequently repeated throughout history, was that they were caused by intercourse with animals. This is evident in the names given to these defects, such as harelip and lion face. The examples of human-animal intercourse resulting in creatures such as centaurs and satyrs were treated as a joke by the Greeks and the Romans. Ancient Judeo-Christians, however, considered animal-human intercourse a serious crime, and individuals suspected of such deviant practices were treated severely. Intercourse with demons is also an old explanation of abnormal development, one too that is not entirely foreign to the current thinking of our civilization—witness the popularity of novels such as *Rosemary's Baby*.

Another theory frequently espoused was **maternal impressions.** Somehow the mental impressions of a pregnant woman were believed to affect the fetus directly; for example, the witnessing of a big fire by a pregnant woman could lead to the formation of a flaming birthmark on her unborn child. This concept has its positive aspects also, and pregnant women around the turn of the century were told that thinking good thoughts and looking at pretty pictures would help ensure that their babies would be healthy and good-looking. This superstition is also quite alive today.

It is important to understand these older concepts concerning the causes of defects because (1) they are still alive, (2) a defective individual is still looked upon with some degree of dread and repulsion, and (3) women who have given birth to a defective child often have strong guilt feelings.

Genetic Factors

What do scientists know about the causes of birth defects? One obvious answer is that some defects must be due to hereditary factors. More than 13 different genes are known that produce 13 different kinds of cleft palate in mice. One in 17 cases of human cleft palate is thought to have a genetic origin, but 16/17 are not. Other defects that have a known hereditary etiology are albinism, sickle cell anemia, phenylketonuria (PKU), and some skeletal abnormalities. The defects caused by chromosomal abnormalities (Chapter 16) also belong in this category. It has been estimated that about 20 percent of human morphological abnormalities are caused primarily by genetic factors. This leaves 80 percent to be accounted for on a nongenetic basis.

Environmental Factors

It was believed, until quite recently, that the human embryo was thoroughly protected from all external influences. Consequently, all defects had to be genetic in origin. This concept was rudely shattered in the 1930s when some pregnant women were treated for cervical cancer with strong x-rays, and their children were born with severe brain damage. The list of substances and physical factors that can affect the development of the mammalian embryo has grown steadily since then. The thalidomide tragedy in Europe, identified in 1961, has led to an increased awareness that the variety of chemicals to which we are exposed may sometimes have less than favorable effects on an individual in the first environment he or she experiences—the uterus.

Agents of these types that can cause defects are known as **environmental** or **external factors,** in distinction to genetic (internal) factors. They are also known as **teratogenic agents** (Greek: teratos, monster; genesis, origin) or sometimes simply as **teratogens.** A list of some agents known to induce birth defects in humans or laboratory animals, or both, is presented in Table 17–1.

The list contains agents in diverse categories: foods, infections, drugs, environmental chemicals, and miscellaneous agents that are difficult to classify. What they have in common is an ability to exert a toxic effect on the developing embryo and fetus. Such toxic effects, if strong enough, can often kill the unborn individual. But less than lethal doses can also have permanent effects on fragile, developing systems; these effects can become manifest as anatomical defects, growth retardation, or functional defects such as mental retardation, endocrine deficiencies, and respiratory problems.

Radiation from x-rays or other sources is a well-known teratogen that can induce a wide spectrum of visible malformations, mental retardation, growth retardation, and embryonic death. The amount of radiation to which the embryo or fetus is exposed is important. Neither the radiation to which we are routinely exposed in everyday living nor a single dental or chest x-ray will harm an embryo or fetus. Radiation from an extensive series of abdominal x-rays may possibly approach harmful levels, however, and the kind and intensity of radiation used for therapeutic treatment will probably be in the range that could be dangerous to the unborn.

Fetal or embryonic **oxygen deficiency** can occur as a consequence of maternal asphyxiation; of twisting of the umbilical cord; of the mother's inha-

Table 17–1 Some Agents Known to Induce Birth Defects or to Have Other Toxic Effects in Humans and Experimental Animals

General Agents	Drugs
Radiation*	Thalidomide*
Oxygen deficiency*	Androgens*
Cigarette smoking*	Cortisone
	Heroin*
Infections and Diseases	Diphenylhydantoin (phenytoin)
Rubella (German measles)*	(anticonvulsant)*
Syphilis	Anticancer agents*
Toxoplasmosis	Aspirin
Cytomegalovirus	Tetracycline
Diabetes*	DES (diethylstilbestrol)*
	Environmental Pollutants
Foods and Additives	Mercury*
Protein deficiency*	Lead*
Vitamin deficiency	Carbon monoxide
Vitamin excesses (particularly Vitamins A and D)	Some pesticides and herbicides
Alcohol (acute)*	
Mineral deficiencies	

*Known to be especially significant in humans.

lation of carbon monoxide; of the mother's ingestion of too much sodium nitrite, a common food additive; or of exposure of the unborn to drugs such as epinephrine or vasopressin, which can constrict the uterine arteries. Oxygen deficiency can induce a wide variety of defects, but fortunately it must reach fairly acute levels before prenatal damage is likely to occur.

Cigarette smoking does not significantly increase the incidence of visible malformations, but it does have a number of unfavorable effects on the unborn, some of which last into childhood. Literally hundreds of studies on many thousands of women and their children clearly document that smoking during pregnancy

1. Substantially reduces the birth weight of offspring.
2. Slightly increases the risk of death of the child before or shortly after birth.
3. Causes slight but persistent decreases in stature.
4. Causes slight but definite decreases in IQ, reading ability, and mathematical skills and increased hyperactivity in school-age children.

Many of these studies have demonstrated a **dose-response relationship;** that is, the heavier the smoking, the greater the effect. Furthermore, cessation of smoking early in pregnancy minimizes the effects. Several studies have also shown that shortly after the mother smokes, decreased breathing (chest) movements and an increased heart rate can be demonstrated in the fetus for more than 1½ hours. The increased heart rate is due to the nicotine, which readily crosses the placenta, but some of the other fetal consequences probably result from the increased carbon monoxide levels in the blood of smokers. Smoking has been shown to have detrimental effects on the structure of the placenta, and this can influence the fetus in many ways.

Rubella, also known as **German measles,** was the first disease that was shown (in 1941) to be a cause of birth defects. This infection produces relatively mild symptoms in the mother, but the virus can cross the placenta and linger in the fetus, where it can induce cataracts, serious damage to the heart and nervous system (brain damage, mental retardation), blindness, and deafness. Maternal infections during the first month of pregnancy are the most dangerous, producing abnormalities in up to 50 percent of the exposed embryos. The cytomegalovirus (Chapter 7) has been shown, on rare occasions, to induce serious brain damage, but no virus approaches rubella in teratogenic potential. The organism that causes syphilis, **Treponema pallidum,** sometimes crosses the placenta. It can cause growth retardation, anemia, deafness, and tooth abnormalities. **Toxoplasmosis** is caused by a parasite that can be transmitted to adults via the eating of uncooked meat or the handling of cat feces. Adults rarely show significant symptoms. Intrauterine infection is, fortunately, uncommon but can produce serious brain damage. Maternal **diabetes,** even if well controlled with insulin, can affect a fetus. The chance of perinatal mortality is increased, the babies tend to be large at birth because of edema and fat deposits, and they sometimes have a defect of the thigh bone.

Protein deficiency and severe **malnutrition** in general can have an unfavorable effect on development, resulting in low birth weight, stunting, and increased prenatal and perinatal mortality. In animals, malnutrition has also been shown to decrease the size of the brain and other organs. Deficiencies of several kinds of vitamins in experimental animals induce a wide variety of defects, ranging from gross brain abnormalities to cleft palate and limb malformations. However, a really severe depletion, not ordinarily encountered in humans, is usually necessary before development is significantly affected. A gross excess of several vitamins has also been shown to be deleterious, particularly the fat-soluble vitamins A and D. Partly for this reason, the amounts of these vitamins that can be added to foods or used in nonprescription vitamin preparations have been greatly reduced in recent years. **Trace mineral deficiencies** are like vitamin deficiencies, inasmuch as they can cause a variety of prenatal problems but extreme levels are required to produce effects. Thus, although iodine deficiency, for example, can induce a form of mental retardation known as cretinism, the practice of iodine supplementation in table salt has virtually eliminated this problem.

Chronic **alcoholism** has been suspected of inducing prenatal problems for a long time, but positive confirmation has been peculiarly elusive. It is only since 1973 that the **fetal alcohol syndrome** (FAS) has been clearly characterized. The major symptoms are pre- and postnatal growth deficiency, an undersized brain, significant mental retardation, and heart problems. Minor changes in facial features and in the creases of the palm of the hand also occur.

About one third to one half of the babies born to chronic alcoholic mothers exhibit some degree of FAS. Although alcoholism is often accompanied by heavy smoking and malnutrition, it is clear now that alcohol per se is the cause of FAS. The significance of the induction of mental retardation by alcohol cannot be overemphasized. It is currently estimated to be the third leading cause of mental retardation in the United States. The governments of several large states, which in the main carry the financial burden of caring for mentally retarded children and adults, and the federal government are sponsoring extensive research into this problem. There is no question that chronic heavy drinking of alcohol during pregnancy can seriously affect the unborn, but exactly what amount constitutes "too much" is not yet completely clear. There is evidence that moderate drinking can produce some degree of FAS in some babies. The possible effects of a single alcoholic binge have not been evaluated. Until these questions are resolved, pregnant women and those who might be pregnant should be judicious in their alcohol consumption.

There was an alarming epidemic in Europe between 1959 and 1961 of babies born with missing or deformed limbs, an otherwise rare defect. The missing and deformed limbs were often associated with heart and kidney anomalies. It did not take too long to correlate these defects with the introduction of a new sedative and antiemetic, called **thalidomide.** This was an unusually devastating teratogen, since very little exposure to this seemingly innocuous drug, especially toward the end of the first month of pregnancy, induced an extremely high frequency of defects. Nothing quite like it has been found since.

Some other drugs are also known to produce prenatal problems. Androgens can effectively masculinize a genetic female. Some synthetic progestins can have the same effect because they are converted in the body to testosterone. **Cortisone** can produce a high frequency of cleft palate in some strains of mice, but humans seem much less sensitive to it. Babies of heroin addicts are born with acute withdrawal symptoms, which must be treated promptly. Although animals exposed to extremely high doses of marijuana extracts can develop defects, there is no evidence that occasional smoking of this material has any harmful effects on the human embryo or fetus. Several forms of **hydantoin,** widely used to control convulsions from epilepsy and concussion, induce growth deficiency, mental retardation, and a characteristic syndrome of facial features and abnormal fingernails in from 10 to 30 percent of the babies prenatally exposed to it. Unfortunately, trimethadione, another widely used anticonvulsant, is also a potent teratogen. Chlorpromazine and a few other tranquilizers at high doses have produced subtle behavioral abnormalities in experimental animals. The selective targets of anticancer drugs are rapidly dividing cells. Embryos have large populations of such cells and hence are also sensitive to these drugs. Normally, a pregnant woman who needs cancer chemotherapy does not receive it until the pregnancy is terminated.

So far, the drugs mentioned are relatively specialized, but the next two are familiar to everyone. Prolonged exposure to ordinary doses of the widely used antibiotic **tetracycline,** either pre- or postnatally, causes a brown staining and weakness in tooth enamel (teeth develop slowly over many years). High doses in laboratory animals can induce other defects. **Aspirin** is an extremely potent teratogen in the rat and monkey. A single heavy dose can induce gross abnormalities of the brain, heart, and other viscera in up to 30 percent of the fetuses. The human embryo seems much less sensitive to aspirin, but exactly how much less we do not know.

Diethylstilbestrol (DES) is in still another category because prenatal exposure to it may cause cancer, which does not become manifest until many years after birth. This drug was frequently given to pregnant women at one time because it seemed to help prevent an incipient abortion. Recently, there has been a mild epidemic of an ordinarily very rare type of vaginal cancer in teenage women. Again, as in the case of thalidomide, a correlation between DES and the cancer was possible because of the unusual nature of the problem. About 1/1000 of the girls prenatally exposed to DES develops cancer. There is also some evidence that a few women and men who were prenatally exposed to DES are having problems with their reproductive systems.

The prenatal effects of environmental pollutants are even more difficult to trace than those of drugs because

1. Exposure to the pollutants is often unnoticed.
2. Detection and identification of pollutants is often a difficult process.

3. Demonstration that very low concentrations of these substances can have unfavorable effects is difficult.

For example, even though a severe outbreak of neurological diseases in fishing villages along Minamata Bay in Japan was first noted in 1955 and was vigorously studied, it was not until seven years later that the cause was identified as **soluble mercury** from industrial waste that was carried by the fish in these waters. The fish would absorb and concentrate the mercury and thus poison the villagers who ate them. In addition to the adults who were poisoned, a number of children were born with mental retardation, limb deformities, and severe brain damage producing palsy-like effects. Pregnant women were often spared of symptoms because the fetuses extracted the mercury from the mother's blood and concentrated it, with damaging effects, in their own brains.

Wheat seed destined for planting and treated with a fungicide containing soluble mercury has been accidentally used for food, which has led to comparable occurrences of mercury poisoning in Sweden and Iraq. Lead has been identified for centuries as a poison with prenatal as well as postnatal effects, yet exposure to lead has been a difficult problem to eliminate. Exposure to carbon monoxide causes oxygen deficiency by preventing the hemoglobin of blood from carrying oxygen. It is a ubiquitous pollutant, and occasionally severe cases of poisoning have resulted in fetal death or brain damage. There has been considerable concern recently about the effects of pesticides, herbicides, and some byproducts of their manufacture (especially dioxin) on prenatal health. Some of these compounds, given at high doses, can induce fetal and neonatal death, growth retardation, and visible defects in experimental animals. Suspected cases in humans have occurred as a result of heavy accidental exposure.

This may seem like an endless litany of visible and invisible dangers to the unborn. A measure of perspective is presented in the next few sections.

Teratogens and Critical Periods of Development

Exposure to a heavy dose of aspirin, thalidomide, or even rubella virus during pregnancy does not guarantee that a prenatal effect will occur. A 1967 study in which pregnant women were interviewed before the birth of their children showed that these women were exposed to an average of 3.7 potentially teratogenic drugs during their pregnancies, yet the incidence of problems in their offspring was only average. Even a laboratory rat or mouse exposed to a potent teratogen can deliver a litter of about 12 pups, of which some will be seriously affected, some will be slightly affected, and some will appear unscathed. The induction of defects by environmental agents, like that by genetic factors, is a matter of odds, and the outcome of a given exposure cannot be predicted with certainty. There are some generalizations that do help to clarify the results of prenatal insults.

Only a very few of the agents mentioned in the preceding sections have highly specific effects—for example, androgens, which affect only the genitalia of genetic females, and cortisone, which produces (in mice) a high incidence of cleft palate and virtually nothing else. The majority of teratogenic agents are nonspecific **toxic insults** to the developing organism and hence are capable of inducing a wide variety of effects. These include death, a spectrum of malformations, growth retardation, and functional deficits. In general, the eventual outcome of a teratogenic insult depends on (1) the intensity of the insult, as indicated by strength of the stimulus and its duration, (2) the period of development in which it occurred, and (3) the degree of susceptibility due to genetic factors and the general state of health of both fetus and mother.

Short, acute exposure to teratogens tends to induce death or malformation. Examples are human exposure to thalidomide and giving a rat a single heavy dose of aspirin. Prolonged exposure to milder noxious insults, such as smoking and malnutrition, is more likely to induce growth retardation and functional deficits.

The period during which an insult is operating is important in determining the outcome. Organs are particularly sensitive to malformation when they are exposed to unfavorable environmental stimuli (1) at the time of their origin or just before this time, (2) during a period of very rapid development, and (3) at the time of onset of function in some organs. In experimental animals it is easy to show that exposure to teratogens before the period of organ development usually has either a lethal effect or none at all, and exposure of older fetuses tends to induce growth retardation or functional deficits because no organs are forming during these periods. But during the period of rapid organ proliferation, teratogens can be especially damaging to any structure in a fragile beginning stage of development. Each organ has its own particular **critical period,** and the early developmental period (two weeks to two months in the human) is critical for the whole organism. Human data as well as animal experiments validate this conclusion. Exposure to the virus of German measles during the first trimester can induce readily apparent birth defects in up to 45 percent of babies, but exposure during the second and third trimesters produces defects in only 5 to 10 percent.

The Important Role of Genetic Background

There is another factor, in addition to the age of the developing fetus, that helps to control the eventual outcome of an exposure to a generalized unfavorable environmental influence: the nature of the total genetic background of the individual. Multifactorial inheritance can affect susceptibility to disease in the prenatal period as well as in the adult (Chapter 16).

An excellent and informative example of this type of effect is cleft palate induced by cortisone, which has been studied extensively. This hormone is one of those rare specifics, a drug that produces a high incidence of cleft palate in mice and little else. To induce cleft palate with cortisone, a researcher injects a pregnant mouse with a small dose (2.5 mg) on each of four successive days, just before, during, and shortly after the palatine bones rotate and fuse to form the hard palate (Chapter 11). If this treatment is applied to one inbred strain of mice, A/Jax, 100 percent of the offspring will be affected. If, on the other hand, this same treatment is given to pregnant mice of another inbred strain, C-57 blacks, only 15 percent of the offspring will have cleft palate. If strain A is crossed with strain C and pregnant hybrids are treated with cortisone, an intermediate sensitivity level of 35 percent is found. If the CA hybrids, which are 35 percent sensitive, are crossed back with strain A mice, which are 100 percent sensitive, and then pregnant individuals are injected with the same regimen of cortisone, *none* of their offspring have cleft palate. By varying the genetic background of these mice, it is possible to obtain strains that are 100 percent sensitive to cortisone, 100 percent resistant to cortisone, or some percentage in between. The genetic constitution of the mother, as well as that of the fetus, has been shown to be significant in this case.

This is an illuminating experiment because it helps to clarify why a certain unfavorable environmental effect will seriously influence the offspring of one mother and have no effect whatever on the offspring of another. It is also important because it points out one of the problems of animal testing for teratogenic effects. Inbred strains of rats and mice are widely used to screen drugs for possible prenatal effects, and there is no way to circumvent the fact that their responses are sometimes variable.

This particular example is unusually dramatic but not unique. Most agents tested will have slightly different effects on various strains of animals. The differences between species can occasionally be considerable. For example, the human embryo is extremely sensitive to the teratogenic effects of thalidomide, but no test animal approaches it in sensitivity. These observations mean that animal experimentation, though very useful and important, cannot completely guarantee the safety of any drug or environmental agent for humans.

In summary, it is apparent that some defects have a direct and simple genetic basis but that the majority are induced by unfavorable environmental influences. These nonspecific insults to the baby can affect the whole organism or almost any organ system, depending on the stage of development at the time the unfavorable influence is present. The genetic constitution of the mother and baby can also affect the response to nonspecific insults.

Birth Defect Prevention and Government Regulation

Much of birth defect research and management is based on the seemingly simple tenet that by identifying agents that can unfavorably affect development, avoidance of these and hence prevention of prenatal damage can be achieved. This concept is more than 100 years old but still frustratingly eludes application in many cases. Even well-known and clearly identifiable teratogens such as radiation and rubella are difficult to control completely. There are problems in detecting and identifying environmental pollutants and other agents that could be prenatal health hazards. Therefore, it is even more difficult to practice avoidance and prevention. Let's examine a hypothetical situation from the standpoints of laboratory findings and human application.

The results of a typical laboratory study of a teratogenic substance, a potent new drug or a toxic pollutant, are shown in Figure 17–1. As one increases the dosage in a logarithmic fashion (i.e., by tenfold increments), the incidence of defects usually proceeds upward in a straight-line relationship from 2 percent to 10 percent or so. As one increases the dosage further, the incidence of defects commonly starts to decrease because lethal effects begin to predominate. Because the rate of spontaneous malformations in most strains of laboratory mice and rats is about 1 percent, it is very difficult to prove with certainty that a very low dose can also slightly increase the possibility of malformation unless thousands of animals are used. This is very rarely done because of the enormous time and expense involved. In the low-dose range, two possibilities can be conjectured: (1) that the dose-response curve continues in a straight line and that even extremely low doses can have a slight but significant effect (a in Fig. 17–1) or (2) that the baby is protected from low doses by either maternal mechanisms or its own well-known abilities to regulate some damage, and hence the curve declines rapidly in this range (b in Fig. 17–1). Both types of situations are known to exist in the few cases in which extensive data in the low-dose range have been gathered.

Government regulatory agencies are then frequently confronted with this

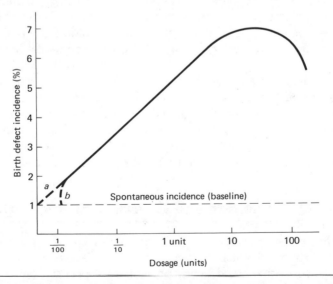

Figure 17–1 A typical dose-response relationship between a teratogen and the incidence of defective development in a laboratory animal.

question: Should a potentially useful drug or chemical (such as a pesticide) be licensed for use even though at moderate or high doses it has been shown to cause prenatal damage in experimental animals? What effect is it likely to have on humans when exposure to only very low levels is likely? Very often at this stage, the benefits to be obtained from the use of this substance have to be considered with respect to its possible risks, the so-called **risk/benefit ratio.** Do the possible benefits of a drug or agricultural chemical far outweigh the slim possibility that it may cause a slight amount of damage to the unborn? Many would immediately say the decision should be clear, that anything that could affect development should be banned, but this is not realistic. Aspirin, though clearly teratogenic in high doses in animals, has not been banned because it is a very useful drug and there is no evidence that it is teratogenic in humans at the doses ordinarily used. Common table salt, sodium chloride, can produce serious prenatal damage if sufficient quantities are injected into mice; however, the amount needed will make the mother quite sick.

These considerations do not mean that a rational approach to the problem of birth defect prevention is impossible, simply that it is a complex situation. As in so many other areas of concern with the environment, clear-cut "yes or no" decisions are not always possible, and some degree of controversy is unavoidable.

Birth Defect Prevention: A Personal Guide

What can you—or your mate—personally do to help ensure that your child is born as perfect and as healthy as possible? Despite the uncertainties described previously, there is a great deal that can be done.

Teenagers should develop and practice good dietary and health habits. Such habits do help to ensure that the final maturation of the reproductive organs, as well as other parts of the body, is not inhibited. This is a good time to be exposed to rubella virus and obtain lifetime immunity against this teratogen. It is also a good time to learn (1) not to take unnecessary medication and (2) not to smoke.

Women should plan to have their children between the ages of 20 and 35, if possible. Plenty of healthy children are born to mothers younger than 20 or older than 35, but the chances of complications occurring to mother or child are slightly higher outside this age bracket. If you have been on the "Pill," stop taking it several months before trying to get pregnant and use another form of contraception. This time allows your endocrine system and reproductive organs a chance to readjust. A rubella vaccine is available but must be taken at least two months before there is any possibility of pregnancy. If you have diabetes, either active or incipient, or if you are taking anticonvulsant medication, consult with a good obstetrician before becoming pregnant.

If you are trying to become pregnant, or are likely to become so, remember that several weeks of the most sensitive period of your baby's life may go unnoticed and that good health care of the child starts before birth. It is important to have a diet that is adequate in protein, vitamins, and minerals. Do not take any medication unless prescribed or at least approved by your obstetrician. Avoid exposure to contagious diseases. You should not smoke or drink, but if you must, keep your drinking confined to less than 1 oz of alcohol (i.e., 2 oz of liquor, 8 oz of wine, or 16 oz of beer) a day and keep your smoking down to less than one pack a day.

Any extensive abdominal x-ray examination (e.g., a gastrointestinal [GI] series) should not be performed except shortly after a normal menstrual period, when the chance of pregnancy is virtually nil. This is a standard procedure in all good x-ray clinics. Exposure to all known teratogens and noxious substances in general should be avoided. Of course, it will not hurt to think good thoughts and look at pretty pictures, too.

Current Trends in Birth Defect Research

One of the primary goals of current research in birth defects is prevention. It is based on the simple assumption that the more teratogenic agents that can be identified and the more that is known about them, the better the position we are in to prevent unnecessary exposure of pregnant women to them. Newer, more sensitive methods of screening substances for teratogenic potential are being developed. Numerous substances are now suspect, based on animal experimentation. More work will be necessary to identify accurately those that are dangerous to humans because of species-specific sensitivity and circumstances of exposure.

Do changes in human birth defect statistics reflect improvement attributable to better prevention practices, or, conversely, a worsening situation because of greater exposure to drugs and toxic industrial chemicals? Defect occurrence is being carefully monitored by epidemiologists (specialists in the gathering of health statistics) with just such a question in mind. The answer at this time is ambiguous because, though the overall level of occurrence of defects has been reasonably constant over the last few decades, there are changes in patterns. For example, the incidence of certain brain and spinal cord malformations is slowly decreasing, but the incidence of heart malformations, specifically ventricular septal defects, is increasing. Both of these are apparently real changes and not due to differences in diagnostic practices. It is almost as if improvements in some areas of prenatal health are being neutralized by additional disturbances in other areas, but more time and study will be necessary to determine what these changes in trends mean.

Prenatal diagnosis of malformations that are not necessarily of genetic origin, specifically spina bifida and related disorders, can be made with amniocentesis. The list of conditions that can be detected this way is growing rapidly. The usefulness of such prenatal diagnosis is, at this time, limited to the recommendation of an abortion when a very serious abnormality is detected.

Much research is also being done on the mechanisms by which teratogenic agents exert their action. This research will help in identifying these agents and in overcoming the problem of species variability. Even more exciting, there is a very real possibility that active treatment can be initiated to cure an embryo or fetus that has been exposed to a teratogenic situation before the damage is irreversible. New terms, such as embryo medicine, fetology, and prenatal medicine, are being coined to encompass these developments in prenatal diagnosis and treatment of birth defects. Even surgery on fetuses within the uterus is beginning to be performed.

There is also a more positive side to birth defect research that goes beyond prevention and treatment. If you have had a chance to examine medieval suits of armor in a museum, you must have been impressed with the fact that the "he man" of medieval Europe was barely more than 5 ft tall. If you have followed basketball you are aware, too, that the average height of the players has increased markedly over the last 20 years. To what can these changes in height be attributed? Genetic alterations alone cannot explain them, particularly the more recent ones. The children of today tend to be taller than their parents because of two factors. The increase in knowledge and better practice of nutrition in the last 50 or more years have ensured that children obtain more of the nutrients they need during their growth periods. The widespread use of antibiotics has virtually eliminated the childhood diseases that were dreaded just 40 years ago. The severe infections that were common then not only killed some youngsters but also produced severe, prolonged illnesses that in many cases stunted growth. Good nutrition and the elimination of some of the factors that tended to stunt growth have produced a new generation of individuals who are, on the average, taller and stronger than their parents and grandparents. Their basic genetic constitution has not changed, but the elimination of unfavorable environmental influences has allowed the maximum potential of the genes that they carry for height and other physical factors to become expressed.

Birth defect research can provide similar insight into the period of prenatal development. As we discover what causes the major shocks that maim and try to avoid them, we also cannot help but prevent situations producing the smaller shocks that can occur before birth. By effectively preventing malformations, we will also be improving fetal health in general and encouraging the optimum utilization of the genetic potential that is already there for the growth and development of such vitally important structures as the brain. This is not an idle dream. The suggestions made previously in the section on personal birth defect prevention already embody this concept, especially with regard to protein nutrition, smoking, and alcohol consumption.

Eugenics as a way of "improving" a group of people does not work, but there is a great deal that can be done toward the fuller utilization and expression of the good genetic constitutions that we have. There is no question that we inhabitants of the planet Earth must have smaller families; therefore, it is feasible to ensure that the offspring we do produce are the best possible.

CHAPTER 18
Family Planning

Family planning means controlling fertility—preventing or promoting pregnancy as desired. The promotion and prevention of pregnancy have been of considerable, often intense, interest throughout human history. A great deal of superstition and folklore is associated with attempts at controlling fertility in virtually every culture. The one thing all these attempts have had in common is failure. Our contemporary era is unique because it is the very first in which family planning can be effectively practiced, usually with relative simplicity.

Promoting Fertility

The immediate concern of many readers of this book is prevention of pregnancy, but for approximately 15 percent of couples in the United States, the inability to have a child is a vexing problem.

Male Infertility

The woman has usually been the focal point in couples with infertility problems, even though about one third of such cases can be attributed to the man. Not all of the reasons for assuming that infertility is a female problem are valid, but it is true that identifying and then assisting a man with marginal fertilizing capacity have been difficult. The only test formerly available was a simple count of total sperm; no attempt was made to distinguish between normal and abnormal sperm. Much progress has been made recently in diagnosing and treating male infertility. There is even a newly developed specialty, **andrology,** which is the male counterpart of gynecology.

The major cause of infertility is a low concentration of healthy sperm. Semen always contains a fair number of immobile and abnormal sperm, but fertility problems can be expected when the concentration of healthy sperm cells dips below 50 percent. These cannot always be detected from a simple sperm count. A thorough semen analysis now includes an examination of fresh semen to determine the concentration and vigor of motile sperm. Other tests are used to distinguish between viable and nonviable sperm.

Modern testing procedures have resulted in the recognition that one of the common causes of male infertility is **varicocele,** a varicosity (swelling) of the spermatic vein (Chapter 4). This condition does not affect sperm concentration, but even a small varicocele can greatly reduce sperm viability for reasons that are not completely understood. Fortunately, this can be corrected by minor surgery.

The numerous contributions of the accessory sex glands to semen have a strong effect on sperm vitality. Semen normally coagulates shortly after ejaculation and then liquefies within 10 minutes. The failure of either of these events to occur properly can indicate a malfunction of these glands. There are chemical tests of semen for various enzymes, minerals, and sugars that can also help in the diagnosis of prostate problems or other disorders within the reproductive tract.

Endocrine imbalances account for about 10 percent of male infertility. The problem can result from (1) a deficiency of pituitary gonadotrophins, either luteinizing hormone (LH) or follicle-stimulating hormone (FSH), (2) inadequate ability of the interstitial cells to synthesize testosterone, (3) hypothyroidism, or (4) other endocrine disturbances.

The results of tests and a physical examination may indicate a regimen of medical or surgical treatment. There are also several procedures that can be tried if no specific treatment is possible for a man with a low count of viable sperm. One method is to avoid intercourse for a full week before the woman's estimated ovulation time. This allows the development of a natural concentration of sperm, which are delivered to the female reproductive tract at a time of optimum fertility. Another technique is to inject the ejaculate (sometimes several collected and concentrated samples) directly into the uterus at the estimated ovulation time, thus avoiding the attrition in viable sperm that occurs in the hostile vaginal environment. This procedure is called **artificial insemination by husband (AIH).**

Artificial insemination by donor (AID) may be attempted if everything else fails. The semen in this case is obtained from an anonymous donor and is injected into or near the woman's cervix. These donors are usually selected to match the husband with respect to physical appearance and blood groups. Sometimes a mixture of semen from several donors is used so that the actual father of the baby cannot be known. The couple is also encouraged to have frequent intercourse around the insemination time to raise the possibility that, should pregnancy occur, the child might be their natural one.

Female Infertility

The reproductive importance of the rhythmical changes in the female reproductive tract and the complexities of fertilization, implantation, and development are discussed in Units One and Two (Fig. 18–1). A normal pregnancy is a rigorous test of the anatomical and physiological adequacy of a woman's reproductive and endocrine systems. A physiological or anatomical problem in any portion of these systems, caused by a congenital factor, accident, or surgery, can result in reduced fertility or frank sterility.

Hormonal imbalance is a major cause of female infertility, responsible for about one third of cases. A quick review of Chapter 8 will show you that there are many places where endocrine problems can originate and disturb reproduction: the hypothalamus, pituitary, ovaries, and even the thyroid gland. These disorders are usually relatively easy to correct with hormone therapy once the cause is determined by diagnostic testing. A careful menstrual history is taken because most of these problems also affect the menstrual cycle. Physical examinations may be done at several different times during a cycle. An important diagnostic step is determining whether ovulation occurs regularly, if at all. Simply determining when ovulation occurs (p. 202) and planning intercourse accordingly to utilize the optimum period of fertilization are often of considerable help. The absence of ovulation can be corrected with a course of fertility drugs. These can be either (1) gonadotrophins with FSH activity, which stimulate the ovaries directly, or (2) the drug clomiphene, which operates via the hypothalamic-pituitary route.

Several relatively minor anatomical problems can interfere with fertility. A malpositioned (retroverted) uterus can prevent pregnancy. This condition can often be helped by surgery. One third of female infertility cases are due to

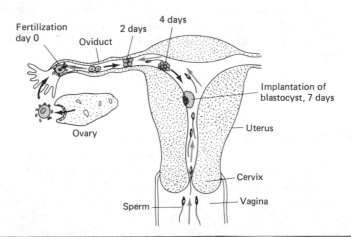

Figure 18–1 A composite diagram of the development of a human embryo during the first week, from fertilization to implantation.

blocked oviducts. These tubes are only ¼ inch in diameter, but their internal epithelium is formed into folds so that the central cavities are of microscopic dimension (Fig. 5–3) and susceptible to injury. A single blocked tube is not important; but bilateral occlusion prevents the passage of both ova and sperm, and the woman is sterile even though the rest of the reproductive system is in excellent condition (Fig. 18–2). The closure is sometimes of congenital origin. Inflammation and scarring, such as from a ruptured appendix or an ascending gonorrheal infection, can also block them. The latter cause is currently commonly observed because of the epidemic of gonorrhea (Chapter 7). Intrauterine devices (p. 207) can also occasionally induce sufficient inflammation to block these tubes. The condition can be diagnosed by x-ray or by introducing a gas (carbon dioxide) into the uterus and determining if it escapes past the ends of the tubes (the Rubin test). Surgery can occasionally correct this condition.

The "test-tube baby" procedure, currently a topic of much notoriety, was originally developed to help women with blocked oviducts have children. The steps in this bypass procedure are (1) to obtain eggs from the ovary of the woman donor by surgery, (2) to fertilize them outside the body by sperm from her mate, (3) to culture them for several days, and (4) to introduce the embryo into the donor's uterus via the cervix (Fig. 18–3). Thus, the first days of existence, which normally occur in the oviducts and uterus (Fig. 18–1), are spent outside the body, and the blocked oviducts are bypassed. This method of *in vitro* fertilization, culture, and subsequent introduction of the embryo into a uterus via the cervix is not new. It has been successfully performed with several kinds of laboratory animals since the 1950s and in pigs and cows since the 1960s.

The embryologist-obstetrician team of R. G. Edwards and P. C. Steptoe of Cambridge, England, pioneered the application of these techniques to hu-

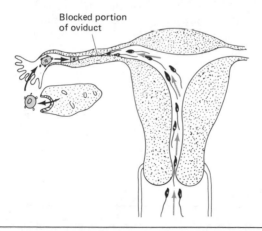

Figure 18–2 The barrier to fertilization presented by blocked oviducts.

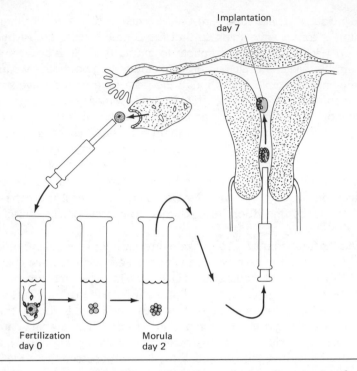

Implantation
day 7

Fertilization
day 0

Morula
day 2

Figure 18–3 Diagram of the bypass procedure, which utilizes *in vitro* fertilization and culture for several days before implantation of the embryo into the uterus.

man eggs and to women with blocked oviducts. Several decades of careful, systematic research by this team finally led, in 1978, to the first successful birth of a "test-tube baby." This spectacular event has received considerable attention in the news media. Clinics utilizing this test-tube technique have been established in Britain, Australia, and the United States. They are currently claiming a 30 percent success rate, about the same rate of embryo survival estimated to occur under natural conditions. It is still too early to predict how common "test-tube babies" will become because of the expense and difficulties inherent in the procedure. Both the technology and the demand for it exist; therefore, it will continue to be perfected.

The "test-tube baby" procedure is just one example of the ways in which the treatment of infertility is progressing. Treatment can now help more than half the women with this problem, and even more may be helped in the future.

Contraception

Contraception is defined as a temporary method of preventing pregnancy, in distinction to sterilization, which is usually permanent. Contraception is a billion-dollar business in the United States and other countries of the world.

There are many pharmaceutical companies whose annual budgets for research and development of new contraceptives are in the million-dollar range. Users of contraceptives have been carefully monitored around the world by both private and governmental agencies. The millions of women on the "Pill" form the most thoroughly studied group of patients in the history of medicine. The styles and trends in contraceptive practice change constantly as a result of these studies. The most popular contraceptive of a decade from now may not even have been discovered yet.

Effectiveness of a Contraceptive

The most important thing to know about a contraceptive is its health hazard potential. The next most important fact is its effectiveness in preventing pregnancy for which there are two measures. The **theoretical effectiveness** is a basic measure of the method itself. It is the optimum success of a method as practiced by reasonably intelligent, careful, well-motivated couples, since the motivation of a couple to avoid pregnancy is particularly important in the success of any contraceptive. All methods are generally more effective in young couples and in older couples who have already had a family of desired size. Contraceptives are least effective in beginners and in couples more interested in spacing rather than in absolute contraception. The more commonly used measure of contraceptive success is **use effectiveness,** the effectiveness of a method as used by an average cross section of a population rather than by a select group. This measure takes human failure as well as method failure into account. Both figures are expressed as **failure rate per 100 woman years** of exposure, that is, how many women become pregnant from a group of 100 who used the method for a full year. The range of values against which the various methods can be compared extends from zero, which would be the rating for a perfect contraceptive, up to 80, which is the pregnancy rate for women who do not use any contraceptive. This latter figure means that from a group of 100 women in the age range of 20 to 35 years who have regular intercourse throughout a full year, 80 will become pregnant. Alternately stated, a woman in her prime period of fertility who has regular, unprotected intercourse has an 80 percent chance of becoming pregnant during any given year.

Some Folk Practices of Contraception

Lactation does inhibit ovulation to a significant degree. This has been noted by many groups of people and may be partly responsible for the custom in some societies of prolonged periods of breast-feeding. It is not, however, a reliable method of contraception. Various kinds of douches following intercourse have been used for centuries. That method is doomed to failure, no matter how effective the spermicide, because sperm can enter the uterus within seconds after ejaculation. Coitus interruptus refers to the withdrawal of the penis from the vagina before ejaculation. It can be used with some degree of success by regular partners. This method is precarious because ejaculation can be precipitous and because some sperm may be present in the fluids that are released prior to orgasm.

Periodic Abstinence

This method of controlling fertility is based on the premise that avoiding intercourse during the time of optimum fertility will reduce the chances of a pregnancy. The success of **periodic abstinence,** often called the **rhythm method,** depends on (1) successfully identifying ovulation and (2) not having intercourse for several days (up to a week) before and after this time. Since there can be problems with both of these factors, this method is helpful in avoiding pregnancy but is not very reliable. It is used primarily for religious reasons, since it is the only method of family planning formally approved by the Roman Catholic Church. It is also used by some couples who prefer a "natural" method based on awareness of a woman's cycles rather than on contraceptive technology. Since the techniques described in the following paragraphs identify the period of optimum fertility, they can also be used to help couples with an infertility problem to achieve pregnancy or simply to maximize the chances of conception at a selected time.

The **calendar method** predicts ovulation time on the basis of a woman's menstrual history. Since ovulation usually occurs 14 days before the next menstrual period (Fig. 18–4), intercourse is avoided for a minimum of four days before and three days after estimated ovulation. This is the fertile, or unsafe, period, and the remainder of the cycle is the safe period. For greater protection, abstinence is often recommended for six days or more before and after estimated ovulation time. But ovulation time cannot always be precisely predicted by this method. When the variabilities of ovulation time, the viability period of the egg (possibly 24 or more hours), and the lifetime of sperm in the female tract (several days) are added together, no time can be considered totally safe. If the calendar method is carefully and systematically used, its failure rate can be as little as 15 to 20 failures per 100 woman years. This much care is seldom exercised, and the usual rate is about 40 per 100 woman years, which makes it just slightly better than no method at all.

There are several newer refinements of the calendar method that improve the accuracy of ovulation detection and hence the success of the method. They are (1) the temperature method, (2) the cervical mucus method, and (3) a combination of all, called the sympto-thermal method. The **temperature method** is based on the fact that in most women there is a slight rise in basal body temperature (BBT) of between 0.4 and 0.8°F, which begins about one or two days after ovulation and lasts for three days. The rise is due to increased levels of progesterone. The temperature is carefully measured at the same time early each morning with a special thermometer. This method determines only that ovulation has occurred and therefore helps to identify the postovulatory safe period. Therefore, for maximum contraceptive effectiveness, intercourse between menstruation and the postovulatory safe period is dis-

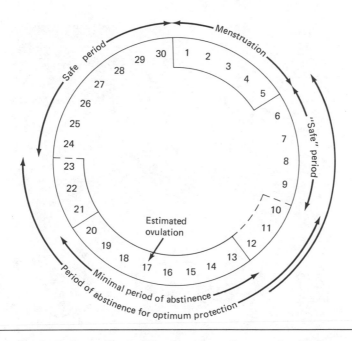

Figure 18—4 A rhythm calendar for a woman with a regular 30-day cycle. In some variations of this method, intercourse is not recommended during the entire preovulatory period.

couraged. The **cervical mucus method** is based on the identification of the cyclical changes in cervical mucus (Fig. 18—5) that occur in response to changing estrogen levels (pp. 55 and 83). There is little secretion of mucus during the infertile days before and after menstruation. The vagina becomes moistened with sparse, sticky mucus before ovulation and with a more abundant, lubricative secretion close to ovulation. To practice this method, a woman must learn to distinguish between sensations of dryness and moistness at the vaginal orifice as well as between different types of mucus. This generally takes several months' training. The failure rate of the cervical mucus method is about 32 to 40 per 100 woman years. The **sympto-thermal method (STM)** uses several indices to identify the fertile period. Changes in cervical mucus and calendar calculations are used to estimate the onset of the fertile period, and alterations in mucus and temperature are utilized to determine its end. The STM method generally takes a three- to five-month training period. Its failure rate can be as little as 3 to 6 per 100 woman years if abstinence is practiced in the entire preovulatory period, but the usual failure rate is reported as 14 to 26 per 100 woman years. The International Federation of Family Life Promotion, the Human Life and Natural Family Planning Foundation (United States), and Roman Catholic Church units offer the requisite instruction for these methods.

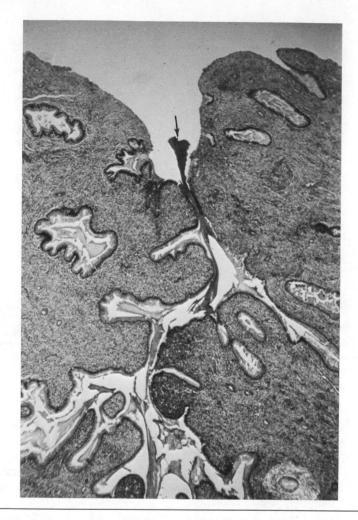

Figure 18–5 A mucous gland of the uterine cervix at or near the ovulatory stage. Note the large, branched gland and dark-staining streaks of mucus in the ducts. Some mucus is being extruded at the orifice of the gland (arrow).

Barrier and Spermicide Methods

The **condom** is a sheath, usually of very thin rubber, that envelops the penis (Fig. 18–6). The condom works by preventing sperm from entering the vagina. It is effective only if it does not have porous areas or any weaknesses that might cause it to split. Many countries set official standards for them, regulating size, thickness, porosity, and bursting strength. The quality of American-made condoms is excellent, although the United States does not officially regulate this product. Modern packaging and promotional methods have added greatly

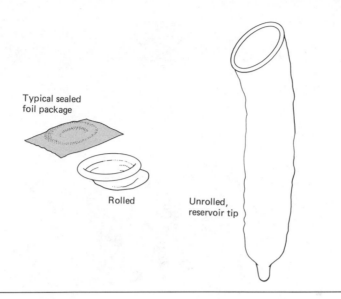

Typical sealed
foil package

Rolled

Unrolled,
reservoir tip

Figure 18–6 Condoms.

to the appeal and availability of condoms. The failure rate is about 7 per 100 woman years. This can be further reduced to less than 3 per 100 woman years when condoms are used in conjunction with a spermicidal foam. The condom is the only contraceptive that also helps somewhat to protect against venereal disease.

The **diaphragm** is a concave disc of thin rubber from 1½ to 3 inches in diameter with a springy edge (Fig. 18–7A). It is designed to fit across the base of the vagina and around the uterine cervix (Fig. 18–7C). Because this is not a sperm-tight fit, diaphragms are always used with a generous quantity of spermicidal jelly, for which the diaphragm serves as a platform. A diaphragm must be fitted by a physician. It must be placed with care before intercourse (Fig. 18–7B) and left for several hours afterward. The failure rate for the diaphragm plus jelly can be as low as 2.6 per 100 woman years if the method is used properly and as high as 10 per 100 woman years if used carelessly.

The **cervical cap** is another type of barrier. It is a small, thimble-shaped cup that fits over the uterine cervix and stays in place by suction (Fig. 18–8). It may be made of metal, rubber, lucite, or flexible polyethylene. It is more difficult to insert than a diaphragm. However, once placed, it can remain for the entire intermenstrual period. They have been used in Europe for a number of years but currently are undergoing tests for effectiveness in the United States. A modification, also undergoing testing at this time, is a custom-molded cervical cap that has a one-way valve for menstrual flows. It can be left in place for several cycles. The failure rate of cervical caps is about the same as that for diaphragms, and some studies, though limited, have shown no failures.

Spermicides immobilize sperm and thus render them incapable of fer-

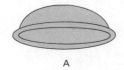

A

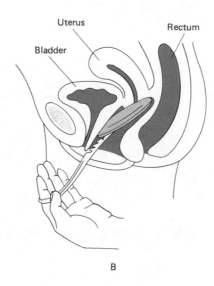

B

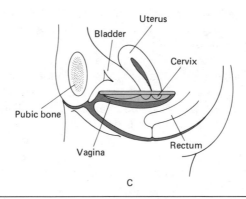

C

Figure 18–7 *A,* An intravaginal diaphragm. *B,* A diaphragm being placed with a plastic introducer. *C,* A properly fitted diaphragm in place. (Courtesy of *Population Reports,* Series H, No. 4, pp. H58 and H63. Population Information Program, George Washington University Medical Center, Washington, D.C., January 1976.)

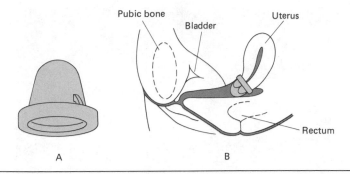

Figure 18–8 *A,* Another type of intravaginal barrier, the cervical cap. *B,* It is shown in place, closely fitting around the base of the cervix. (Courtesy of *Population Reports,* Series H, No. 4, p. H70. Population Information Program, George Washington University Medical Center, Washington, D.C., January 1976.)

tilizing an egg. These agents are in the form of creams, jellies, or foams that are introduced into the vagina before intercourse. They are packaged in tubes, suppositories, or aerosol cans. Foams can also act as barriers to a limited degree because as they expand they tend to fill the vagina. The effectiveness of spermicides is constantly being improved. Newer methods of packaging them into one-dose, purse-sized units are adding to their convenience. The failure rate, when they are used alone, is between 5 and 10 per 100 woman years. The combination of condom plus foam (Fig. 18–9) or diaphragm plus jelly can increase this effectiveness to 2 to 3 failures per 100 woman years.

Recently, there has been a marked increase in the popularity of condoms and diaphragms because they have no side effects. This represents a "back-to-basics" trend in contraceptives, in part stimulated by concern about some of the more serious side effects of the "Pill" and IUDS. In conjunction with spermicides, diaphragms and condoms do offer inexpensive, safe, and effective contraception. Their big disadvantage is that they must be applied before intercourse, and, consequently, in haste they may not be inserted properly.

A new variation on a very old theme is spermicide-soaked sponges. The powder puff type is worn like a diaphragm; another type is cup shaped and fits over the cervix. These sponges are designed to be worn for two-day intervals; they are then taken out, rinsed, and resoaked in spermicide. Currently undergoing field trials, these sponges have reported failure rates of less than 5 per 100 woman years.

Intrauterine Devices (IUDs)

Intrauterine devices are variously shaped loops or discs of flexible plastic or metal (Fig. 18–10) that are inserted into the uterus, where they may remain for several years. They have been very popular because of their convenience and effectiveness, with a failure rate ranging from 1 to 6 per 100 woman years.

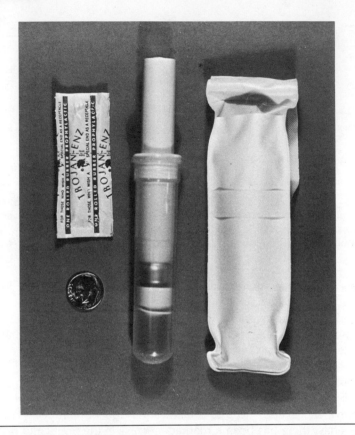

Figure 18–9 A packaged condom and a purse-sized tube of spermicidal foam in an aerosol/applicator package. This is an inexpensive and effective contraceptive combination in a compact and convenient package that can be carried in pocket or purse.

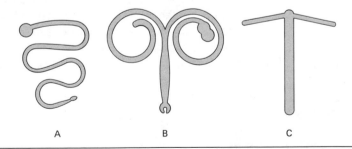

A   B   C

Figure 18–10 Several types of IUDs. *A*, Lippes Loop. *B*, Saf-T-Coil. *C*, Progestasert, which has a vertical section that contains progesterone in a slow-release system. A similar model has a vertical section wrapped with copper. Both copper and hormone enhance the efficiency of the IUD.

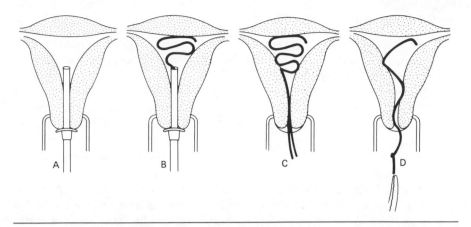

Figure 18–11 Diagrams showing placement and removal of an IUD (Lippes Loop). *A,* The tip of the introducer is passed through the vagina and uterine cervix. *B,* The collapsed IUD is gently extruded. *C,* An IUD correctly in place with strings extending into the vagina. *D,* An IUD being removed.

IUDs are somewhat different from the other contraceptives discussed so far because they do not prevent conception but somehow preclude the implantation of a blastocyst into the wall of the uterus. This process is not fully understood. Some newer types of IUDs contain copper wire or slow-release hormones, which improves their effectiveness.

An IUD is placed by first collapsing it into a soda straw–like device called an introducer. This device is then inserted through the cervix, and the IUD is delivered into the lumen of the uterus by a plunger (Fig. 18–11A). Once released, the IUD springs back to its natural state (Fig. 18–11B). An IUD is placed with a portion of string left extending from the cervix into the vagina. This string not only facilitates removal (Fig. 18–11D) but also allows the user to check for retention of the device, since IUDs are occasionally expelled spontaneously (about 10 percent during the first year of use). Other problems that can occur in some women are uterine cramps and prolonged menstrual bleeding (another 10 percent). These problems are more likely to occur in women who have not had any children than in those who have. Occasionally, placement of an IUD can trigger an infection or even lead to a perforated uterus. An IUD can also sometimes cause an inflammation that ascends into the oviducts, where it can cause a blockage and infertility. A rare but potentially serious side effect of IUDs is that if a woman should become pregnant while wearing one, there is approximately a 1 in 25 chance that the pregnancy may be ectopic (p. 164). Nevertheless, IUDs are conveniently used by millions of women around the world, and IUD technology is constantly being improved with respect to design and fitting and placement methods.

Oral Contraceptives: The "Pill"

The "Pill" is a combination of synthetic estrogens and progestins taken in such a sequence that ovulation is inhibited but menstruation is not. The elevated blood levels of the hormones produced by taking the Pill affect the hypothalamic feedback mechanism (Chapter 8), and the typical midcycle surges in LH and FSH do not occur (Fig. 18–12). Therefore, no ovum matures, no ovulation occurs, and no pregnancy is possible. Theoretically, the Pill should be nearly perfect in preventing pregnancy, but in actual usage, it has a failure rate of 0.3 to 1.0 per 100 woman years. It is still the most effective contraceptive currently available.

Most oral contraceptives are dispensed in sets of 21 or 28 pills in a container that numbers each tablet (Fig. 18–13). Every pill of a set of 21 contains hormones. One pill is taken each day, beginning on the fifth day of a cycle. No pills are taken for the other seven days of the cycle. When there are 28 tablets in a set, 21 of them contain hormones, and the other 7 are blanks, which are included to facilitate the routine, so that one pill is taken every day. The blood levels of hormones drop after the twenty-first day in either case. This drop in hormones triggers a menstrual period, which starts about three days later, just as a similar drop initiates a naturally occurring menstrual cycle (Chapter 8). It is for this reason that contraceptive pills can be used for regulating cycles as well as for preventing pregnancy.

Oral contraceptives are available in two varieties: the combined type and the minipill. Each of the 21 pills of the combined variety contains estrogen and

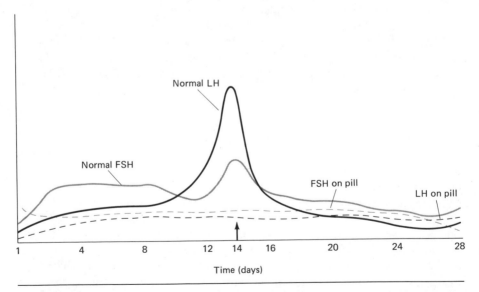

Figure 18–12 Blood levels of FSH and LH at different phases of a typical cycle in a normal woman and a woman on the "pill."

Figure 18–13   Typical package of oral contraceptives.

progesterone. The minipill, which is taken daily, contains only a small dose of synthetic progesterone.

Oral contraceptives are known by a great many trade names, contain a variety of different synthetic hormones, and have several different formulations. The varieties and formulations are constantly changing as research and clinical experience indicate. One long-term trend, for example, has been to decrease steadily the hormone level in the pills. The latest type is a phased pill in which the dose of progesterone is increased in a stepwise pattern during the 21 days of pill taking. Changes in formulation will continue because oral contraceptives are a highly competitive billion-dollar business closely watched by pharmaceutical companies, physicians, and users alike. There were 50 million users of the pill in 1973. Twenty percent of women between the ages of 16 and 44 in the United States, 25 percent of Australian, Canadian, and West German women, and 37 percent of the women in the Netherlands were using it that year.

These astonishingly high figures testify to the many advantages of the Pill. It is a very effective contraceptive. It is neat and does not require any precoital manipulation of the genitalia or interfere in any way with the spontaneity of sexual activity. It is also, for most women, a safe contraceptive to use. However, for some women, it presents problems. The Pill contains potent hormones that affect many organs, and numerous side effects are possible. These may be serious—for example, blood clots (thrombosis), gallbladder disease, and high blood pressure—or minor—for instance, nausea, fluid retention, mood changes, headaches, growth of facial hair, and a tendency toward candidiasis (Chapter 7). The serious side effects are more likely to occur in women older than 35. As many as 40 percent of women on the Pill develop some side effects, but usually they are minor. Careful matching of the woman to the formulation of the specific contraceptive pill, taking into consideration age, health, and other factors, minimizes the possibility of side effects. Most doctors and clinics do this. Most doctors also request a follow-up examination after the woman has been taking the Pill for 6 to 12 weeks. Any side effects that appear can frequently be eliminated by changing the formulation slightly. Most of the side effects are due to estrogen. A woman who persists in having physical problems from the combined pills and still wishes to continue with oral contraceptives is a good candidate for the minipill. The theoretical effectiveness of progesterone-only preparations is 1 to 1.5 failures per 100 woman years, just slightly lower than that of the combined variety.

The Pill is not, at present, a drug that can be considered innocuous. Prolonged usage, for instance, can result in hypersuppression of the hypothalamus and a failure to ovulate for some time after the use of the Pill stops. It is for these and other reasons that this contraceptive method must be used under the supervision of a physician.

A "one-a-month" or "morning-after" pill is a desirable goal of contraceptive research. One postcoital contraceptive is already available. A moderately strong dose of estrogen, given either by mouth or by injection for 4 consecutive days, can effectively prevent pregnancy if administered within 48 hours after intercourse. **Diethylstilbestrol (DES)** is commonly used for this purpose. This is a potent treatment with unpleasant side effects such as nausea, vomiting, and tenderness of the breasts. It is used only for emergencies such as rape or contraceptive failure at a time when pregnancy would be traumatic. "The Shot" is an injection of progesterone in slow-release form that gives continuous contraceptive protection for three months. It is used in several countries. But as recently as 1978, the Food and Drug Administration (FDA) has rejected it for use in the United States because of unpleasant side effects such as irregular or absent menstruation.

Contraception in the Near Future

Because of the great popularity and importance of contraceptives, new developments in the field are covered widely in newspapers and magazines. It is sometimes difficult to tell from these articles whether a method is (1) still undergoing laboratory testing (Phase I) and is years from the market, (2) at present undergoing limited (Phase II) or extensive (Phase III) field trials on humans under carefully controlled circumstances, or (3) currently available to the general public. These are the normal evolutionary stages of a new medical product, with federal licensing required at all but the most preliminary stages. All of the products discussed in the following paragraphs are still in one of the developmental stages.

Improvements in existing forms of contraception can certainly be expected. New spermicides and improved carriers for them that will increase their effectiveness and convenience are being developed. Various ways to introduce the hormones of the Pill in a slow-release form, either as a capsule implanted under the skin or as a small intravaginal ring, are being tested. A single placement of either device would probably be effective for several months.

Some rather exotic new forms of contraception are based on immune mechanisms. These are still very much in the trial stage. It is possible to immunize a woman against her mate's sperm, which is something that can occasionally happen naturally. Whether spontaneous or induced, such immunity results in infertility because the sperm clump together and are thus incapable of fertilizing an egg. An antipregnancy vaccine, which immunizes a woman against the hormone human chorionic gonadotrophin (hCG), has been tried. The major problem with either of these methods, apart from the side effects, is that they are difficult to reverse.

Another new type of chemical contraceptive, currently in Phase II trials, interferes with messages from the hypothalamus. Both luteinizing hormone–releasing factor (LH-RF) and LH-RF–like compounds have been synthesized. These substances, given by injection or as a nasal spray at inappropriate times of the cycle (p. 83), effectively inhibit ovulation with few side effects. These chemicals may possibly be effective postcoital and male contraceptives.

An important new group of drugs called prostaglandins are undergoing various levels of testing. They are naturally present in semen and other tissues, where they have a hormone-like activity. Some of these prostaglandins will be used to treat blood pressure, asthma, and ulcers. Two of them are of particular significance for reproductive biology. They are the forms known as PGE_2 and $PGF_{2\alpha}$. The main effect of PGE_2 is to stimulate uterine contraction. It has already been used to bring on labor in problematic cases and to induce abortion without surgery during the third to fifth months of pregnancy. IUDs containing this substance in a slow-release form are being tested. $PGF_{2\alpha}$ adds a new dimension to birth control. It not only causes uterine contraction but also interferes with corpus luteum formation and thus inhibits the progesterone synthesis that is necessary for the maintenance of pregnancy. $PGF_{2\alpha}$ has a double-barreled contraceptive effect. It (1) irritates uterine muscles at the delicate time of implantation, preventing that step from occurring, and (2) inhibits the formation of the corpus luteum, which will stop a blastocyst from developing any further even should implantation have taken place. Strictly speaking, it is not a contraceptive because it does not prevent conception, but it does prevent pregnancy by interfering with implantation, as does an IUD. $PGF_{2\alpha}$ or one of its chemical cousins may soon be a safe and convenient "once-a-month" pill, to be taken about the time a menstrual period is due. Some forms of prostaglandins are currently undergoing Phase II trials.

The actions of the chemical contraceptives of today and tomorrow that are reviewed in this chapter illustrate the complexity of the physiology of the female reproductive process. The numerous points at which this process is vulnerable present ample targets for the scientists developing new contraceptives.

Why is the woman the primary target of most contraceptive products? Why has there been no development of a male pill? Several have been tried, but it seems much easier to interfere with the release of a single egg once a month than to stop the development of millions of sperm taking place in a day. Paradoxically, high daily doses of testosterone will induce temporary sterility by inhibiting spermatogenesis, but the large amounts of hormone required for this carry a high risk of liver damage. There are drugs that inhibit the synthesis of interstitial cell–stimulating hormone (ICSH) by the pituitary. These eventually inhibit testosterone formation by the testis and stop sperm formation. Unfortunately, because of the lack of testosterone, the treatment also induces loss of libido, a self-defeating side effect of any contraceptive. High concentrations of estrogens can also interfere with sperm formation, but these have strong feminizing effects, which are not desirable. There are several drugs that kill or inhibit the stem cells of the testes, but none of these seem even reasonably safe to use at this time. One of the dangers is a great potential for genetic damage that may not be apparent for several generations. A group of chemicals related to insecticides induce sterility by killing maturing sperm cells, but, again, with no margin of safety when they are tested on rats. There are other substances known that affect male fertility by immobilizing sperm, breaking their tails off, or otherwise rendering them incapable of successfully fertilizing an egg, but none of these is safe enough to use at this time. Gossypol, a substance found in cottonseed, is being extensively tested as a male contraceptive in China. This too has numerous side effects, such as fatigue, decreased libido, blood changes, and incomplete recovery. Gossypol is not being tested in the West. No other male contraceptive is currently undergoing Phase II trials. The immediate future of this area of contraceptive technology is uncertain, despite considerable interest.

Surgical Sterilization

Female Sterilization

Elective, voluntary sterilization as a method of limiting family size is a relatively new phenomenon and is rapidly becoming more popular. The number of people around the world who are electing sterilization has increased from about 3 million in 1950 to 65 million in 1975.

The early history of attempts to sterilize women is long and bizarre, consisting usually of cruel, mutilating, and dangerous procedures. Since the discoveries of anesthetic and antiseptic techniques around the turn of the century, more than 100 different surgical ways to sterilize women have been described. The removal of a uterus or both ovaries for significant medical reasons will obviously also sterilize the patient. Voluntary sterilization, however, is usually synonymous with tying off and severing the oviducts, a procedure called tubal ligation (Fig. 18–14). This operation is usually done through an abdominal incision with the patient under general anesthetic; it is followed by a short stay in the hospital.

The great popularity of tubal ligation has brought about the development of many procedures that are easier and safer for the patient. In underdeveloped countries, some of these can be done with minimal hospital facilities. Sometimes the tubes are simply clipped or electrically burned (fulguration) instead of being severed. Instruments are now available that require only a small abdominal or suprapubic incision ("Band-Aid" surgery). It is possible to use a local rather than a general anesthetic for those procedures, which reduces the need for postoperative hospitalization. Sterilization surgery is sometimes performed through the vagina. Fulguration can also be done by inserting instruments into the oviducts through the vagina and uterus, a procedure that requires no incision. Still in the experimental stage are methods for blocking the tubes with various substances delivered by the intrauterine route. Blockage can be produced by chemicals that cause inflammation and scarring, by adhesions,

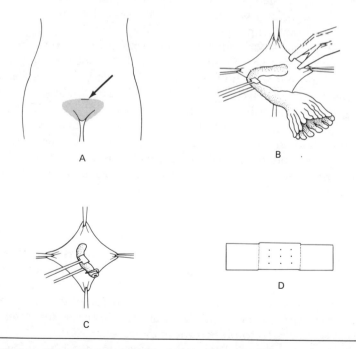

A

B

C

D

Figure 18–14 Diagram of a tubal ligation. *A,* The 1-inch abdominal incision below the line of pubic hair. *B,* The oviduct is brought out through the incision, ready for clipping and tying. *C,* The remainder of the oviduct is replaced. *D,* The "Band-Aid" bandage of the incision.

or by polymerizing plastics that are placed while in the liquid state and allowed to harden in the tubes.

The possibility of reversing surgical sterilization depends on the procedure used. It is sometimes impossible and sometimes has limited success. Attempts to produce temporary sterilization by plugging the tubes with a plastic insert have met with some success. These plugs are placed either by the intrauterine route or by an abdominal incision.

Male Sterilization

Male sterilization can be accomplished either by **castration** or by **vasectomy**. Removal of the testes is a drastic and permanent form of sterilization because it eliminates the source of the male hormone as well as the sperm. It is sometimes necessary in cases of accident or cancer and has at times been meted out as a punishment. Curiously, vasectomy, a tying off of the vas deferens, was not done until 1890. The operation was originally developed to prevent an infection from descending to the testes after prostate surgery. It was not used for sterilization until early in the twentieth century. Vasectomy is a simple, inexpensive, safe, and effective method of fertility control. It is simpler and easier on the patient than any form of female sterilization used today. Approximately ½ million vasectomies are performed annually in the United States and over 1 million per year in India.

The vasectomy is performed through a small incision on the scrotum, just above the testes. The vas deferens of each side is located and freed from artery, veins, and nerves. The vasa are then cut, tied, and electrocoagulated and/or sealed with metal clips (Fig. 18–15). The procedure is usually performed with a local anesthetic and can be done within 10 to 15 minutes in a doctor's office, a clinic, or a mobile unit. The patient can leave after a short rest of ½ to 1 hour and is usually able to resume normal activity the next day.

In physically and emotionally healthy men, vasectomy does not affect the male hormone balance, libido, erectile capacity, ejaculation, or orgasm. There are two problems, however, with this procedure. The first is that complete sterility does not occur until the sperm already stored in the reproductive system are ejaculated, which may take several days or even weeks. Another form of contraception must be used until a sperm count confirms sterility. The other problem with vasectomy is that the probability of successfully reversing it is limited. Restorative surgery consists of suturing the cut ends together and trying to keep the lumen open, which is the most difficult aspect of the procedure because the lumen is so small (Fig. 4–11). Restoration of the muscular motility of the vas deferens is also a problem. Functional success (as determined by pregnancy) occurs in 18 to 60 percent of the cases.

The demand for surgical reversal of sterility in men is very low, about 1 per 1000 vasectomies in the United States, Korea, and India. However, the popularity of vasectomy would undoubtedly increase if the successful reversal of the procedure were easier and more certain. Considerable research and

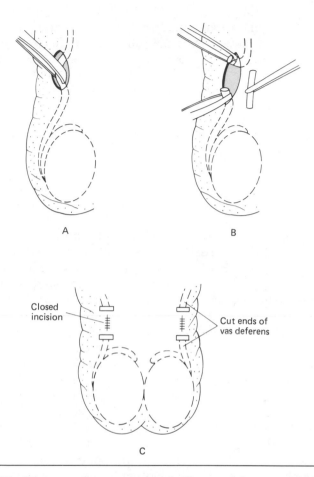

Figure 18–15 Diagram of a vasectomy. *A,* The vas deferens is lifted out through a small incision high on the scrotum. *B,* A small piece of the vas deferens is cut off, and the cut ends are sewn together or clipped. *C,* The ends are replaced into the scrotum, and the incisions are closed.

development toward this goal is being done throughout the world. One alternative available at present is to store a number of sperm samples in the deep-freeze of a sperm bank before a vasectomy, to be used when desired to induce pregnancy by artificial insemination. This is a reasonably successful procedure that is still being perfected (the costs in the United States are $100 for initial processing and about $25 per year for maintenance). Truly reversible vasectomies may be performed by either (1) occluding the vas deferens with a plug or thread that can eventually be removed with relative simplicity or (2) inserting a reversible valve into the vas deferens. Occlusive devices do not work very well because the pressure of the builtup sperm tends either to dislodge the plug or to force the sperm around it. There are two different types of valves that seem to have promise. One is a screw-type valve that is inserted between the two cut ends of the vas deferens and can be placed (by a surgeon) into an open or closed position. The other is a three-component device that is inserted into an incision in the vas deferens. The two end pieces are hollow, and the center piece, the interchangeable one, is either a solid plug or a hollow tube. Both devices require considerable surgical skill to install, but reversing the procedure is relatively simple. Another system utilizes a magnetic on-off valve that can be switched without additional surgery. All of these devices are now undergoing clinical trials on human volunteers.

Abortion

A pregnancy that is terminated for any reason before the fetus weighs 500 grams (1.1 lb) is called an **abortion**. This weight was chosen because there is no chance that a fetus below this weight can survive after birth. A fetus reaches this weight when it is about four and a half to five months old. Abortions can occur spontaneously or can be induced artificially by the pregnant woman or by a physician or other party. Induced abortions can be either legal or illegal. The liberalization of abortion laws in the United States, Britain, and most other parts of the world has made legal abortions under medically supervised and safe conditions readily available, whether for potentially dangerous or simply undesired pregnancies. Self-induced and illegal abortions are still sometimes performed because of ignorance, a desire for privacy, or economic reasons.

The method used for terminating a pregnancy depends to a large degree on the stage of fetal development. It is easiest to end an early pregnancy. Up to three months, the contents of the uterus can be removed by passing a tube through the cervix and gently suctioning the contents, a process called **vacuum aspiration**. Aspiration for a possible very early pregnancy is called **menstrual extraction**. A very small plastic tube (Karman cannula) is inserted through the cervix, and a gentle vacuum is applied. This simple procedure can be used for a menstrual period that is late by a week or two, which may be a pregnancy that is too early to be diagnosed or just a late period. It has been used by "do-it-yourselfers" simply to eliminate the inconvenience of a menstrual period—hence the name. Regular use for this purpose is no longer considered advisable because of the danger of infection or injury to the cerivx and uterus.

Another method of abortion that can be used during the first trimester is a D & C, or **dilatation and curettage**. With the patient under a general anesthetic, the cervix is gently and gradually dilated, and the contents of the uterus are scraped out with a tool called a curette.

Neither aspiration nor a D & C is suitable for the second trimester of pregnancy. During this period the following procedure is commonly used. Some amniotic fluid is removed through the abdominal wall and is replaced slowly with a strong solution of table salt or urea. The fetus is usually expelled 24 to 48 hours later. Prostaglandins are being used more frequently to induce second trimester abortions. The cervix is slowly dilated by one of several methods, and the prostaglandin is applied either by vaginal suppository or by injection into the amnion or maternal muscles. The abortion follows in about 24 to 48 hours. Surgical removal of the fetus by cutting through the uterus (hysterotomy) used to be the only procedure for terminating midtrimester pregnancies, but it is usually not performed now unless a sterilization or other surgical procedure is also planned.

Any abortion carries some risks. The major ones are (1) uterine bleeding from perforation or incomplete removal of the placenta and (2) infection. The first usually happens very shortly after the abortion. The latter may take several days to develop. The risk of these complications is minimal if the abortion is performed under good hospital or clinical conditions. Prompt diagnosis and care further reduce their significance. The risks, however, can be considerable if the procedure is self-induced or performed in an illegal abortionist's office.

Abortions are still considered major surgical procedures. Even though they are readily available today, they should be reserved only for emergencies and should not be considered either a routine method of family planning or a substitute for contraception.

Addendum Since mid-1982, when this book went into production, several studies have been published which report that women who have used oral contraceptives for a year or more gain some positive health benefits. These extend beyond the purposes for which the Pill is primarily prescribed, namely contraception and treatment of menstrual disorders. The risk of developing ovarian and endometrial cancer is reduced by one half, a protective effect that persists for several years after use of the Pill is discontinued. Women taking the Pill are less likely to develop pelvic inflammatory disease and benign breast disease. Contrary to previous concern, the use of oral contraceptives does not increase a woman's risk of developing breast cancer or appear to influence her risk of developing cervical cancer. However, the increased risk of circulatory system disease for oral contraceptive users who are over 35, particularly those who also smoke, has been confirmed.

CHAPTER 19

Human Sexuality

Sexuality, as the word is currently used, encompasses numerous physical, emotional, and sociological facets of sexual activity. The biological aspects of sexuality have been discussed with some degree of thoroughness in this book. But humans are more than just the product of genes, anatomy, and hormones; we are also feeling, thinking, cultural individuals. Sexuality cannot merely be equated with reproduction; it also refers to a powerful and compelling emotion that pervades many aspects of our daily lives. The purpose of this chapter is not to provide an in-depth analysis of this subject but simply to discuss briefly a few psychosocial aspects of sexuality for some measure of perspective.

Nature of the Sex Drive

The sex drive is recognizably a very strong force in our lives. Historic and prehistoric groups of people who did not have a powerful drive to reproduce simply did not leave successors. Mating thus became a strong innate urge, a part of our genetic heritage. The biological components of the sex drive are the hormones that stimulate libido as well as some of the arrangements of nerve cells deep in several crevices of the brain.

Some aspects of the sex drive are psychological. Learning that sexual activity provides pleasure is significant because the pursuit of pleasure is a very potent drive. The use of sexual activity to acquire social prestige or for financial gain is also a learned response. Most people have a conscious desire to reproduce their own kind. It is, after all, one way to achieve a degree of immortality on this earth. Historically, considerable group pressure in most societies has been exerted on young individuals to have children.

Hence some of the components of the sex drive can be classified as innate urges and some as learned behavior. All are intricately woven together in the complex, richly textured fabric that is the personality of an adult. Scientists may disagree about the relative importance of genetic (i.e., innate) and learned influences on behavior, but none would seriously deny that both are significant in sexual behavior.

Development of Personal Sexuality

We have already discussed numerous features of sexual development. The combination of sex chromosomes that results when an egg is fertilized by a sperm determines an individual's genetic sex—either XX female or XY male. An early critical expression of the genes on these and other chromosomes occurs when **gonadal sex** is established (five to six weeks), because gonads play a role in the development of **anatomical sex** (two to three months). The hormones produced (or not produced) by the gonads during the fetal and perinatal periods also permanently affect the hypothalamus and undoubtedly other parts of the brain as well. If all proceeds normally, at birth these phases of sexual development are in accord, and the individual is of definitive sex.

The first stage of psychosexual development involves the gradual establishment of **gender identity,** the personal awareness of being male or female. This process starts at birth with the public recognition of the nature of the external genitalia. Those babies with male genitalia are boys, and those with female genitalia are girls. The choice of a name and the use of the male or female pronoun automatically follow this recognition and start to make daily impressions on the child. Numerous social conventions and boy-girl differences in clothing, toys, and tasks continuously reinforce the initial distinction. Gradually, the child perceives that he or she is of one gender and that some children are of another, even though the basis for this distinction may not be apparent to the child until later.

The establishment of gender identity, the conception of oneself as being male, female, or ambivalent, progressively leads to the next phase of psychosexual differentiation, the development of **gender role.** This is the public expression of an individual's gender identity, that is, what a person does to indicate to others that he or she is male or female. The processes of establishing gender identity and gender role start in infancy and childhood, usually reach a definitive state during puberty, and continue to be refined into adulthood.

Some abnormalities of sexual development have already been discussed, cases in which genetic sex, anatomical sex, and hormonal sex either are not in agreement or are ambiguous (Chapter 13). The treatment of an individual with such abnormalities generally consists of deciding, at birth, the most logical sex for that child and then naming and raising him or her accordingly. Eventually, surgery or sex hormone treatment or both can enhance this decision. These individuals have a healthy psychological development if they are treated with consistency. An attempt to change their gender identity, even as early as age two, can lead to problems. This fact, along with other observations, indicates that gender identity is firmly established by the age of one and a half to three years.

A rare anomaly, but one that is frequently the source of considerable attention, is **transsexualism.** It can be defined as a discordance between anatomical sex and gender identity. The individual genuinely feels that despite his or her anatomy and being raised as one gender, he or she is really of the opposite sex. This condition is more common in men (who feel they are women) than in women (who feel they are men) by a ratio of 4:1. Transsexuals may wear clothes appropriate for the sex they desire, be treated with appropriate hormones, and even undergo extensive surgery to complete the transformation.

Gender identity and gender role are not the only significant factors in the development of sexuality. Love can be separated from sexual conduct, but for most people and in most situations, love is closely associated with sexual behavior. The development of this emotion in an individual's personal history normally proceeds in stages. The first love bond the child forms is for the mother, and it is a response to the comfort of bodily contact and nursing as well as a reflection of the mother's attitude. This phase of close attachment and dependence for life functions can be shared with the father and establishes a child's basic security. Once this occurs, progressive independence—psychological weaning—can evolve and prepare the child for the next phase, which is peer love. This period starts around age three and lasts into puberty, when affection between peers normally develops into heterosexual love. Substages and phases of heterosexual love can be identified—for example, the sequence of dating, kissing, petting, genital stimulation, and eventually intercourse. The uniquely human emotion of romantic love can occur at some time during this growth and is the natural culmination of the love bonds that began forming immediately after birth.

Varieties of Sexual Activity

The sex drive can be strong at times and negligible at others. It can be satisfied by holding hands with a friend or by nothing less than vigorous lovemaking. Sexual feelings can also be ignored or sublimated into physical activity or some other type of effort, which is a common occurrence.

Sexual activity is no longer strictly synonymous with heterosexual intercourse. Autosexual and homosexual activities are more openly recognized as acceptable ways of meeting sexual needs and desires. Even celibacy has been rediscovered and proposed as a suitable option for some periods of life.

Autosexual Activities

Most people prefer a partner, but there are occasions or periods of time when a partner is not available or simply not desired. Autosexual forms of activity include fantasy, dreams, and masturbation. Erotic daydreams are very common, probably nearly universal in occurrence. They may be of a fleeting quality or of considerable intensity. Fantasies may be explicitly sexual (especially in men) or have a strong emotional and romantic theme (more characteristic of women). They may simply be an expression of wish fulfillment and thus be a readily available source of pleasure that can provide a mild degree of satisfaction in the absence of an opportunity for a fuller expression of sexual desires. Fantasies may recall (and frequently embellish on) a past experience. They may also center on future anticipations and thus be useful as a preparation for new situations. Erotic fantasies serve many useful and healthy functions.

Sexual dreams are also of nearly universal occurrence. The erotic content of dreams may not be immediately apparent because it is frequently camouflaged by symbolism. Sexual dreams occasionally result in orgasm in both women and men. Such a nocturnal orgasm usually causes the person to awaken and feel some sexual pleasure but may also be associated with a degree of discomfort. There is some evidence that nocturnal orgasms may act as natural compensatory outlets for sexual needs, since they are more likely to occur in the absence of other sexual activity. However, they do not seem to be a very effective method for satisfying sexual needs.

Masturbation is the physical stimulation (usually manual) of the genitalia and other erogenous zones, leading to an orgasm. It is most often a form of self-arousal but is sometimes done with a partner or in groups. In men it primarily consists of penile stimulation. In women it involves clitoral stimulation, which may be supplemented by stimulation of the breasts, mons, vulva, and vaginal orifice. The stages of bodily responses during masturbation—that is, excitement, plateau, orgasm, and resolution—are the same as in intercourse.

Masturbation is a well-documented occurrence in innumerable cultures, ancient as well as contemporary. In the United States, most men and many women admit to masturbating at some time in their lives. In college students, masturbation may be infrequent or may occur weekly or daily. A significant number of married people of both sexes masturbate occasionally, and the practice is not unknown in senior citizens. Young people masturbate to learn about their own sexual responses and to relieve sexual tension. Adults also masturbate to supplement other forms of sexual activity, to experience fantasies, to help contend with loneliness, or simply to relax.

For more than 200 years in our Western culture, masturbation has been vigorously condemned for numerous physical and moral reasons. This practice of "self-abuse" was held to be responsible for insanity, heart murmurs, nosebleeds, painful menstruation, skin problems, and many other conditions. Some of this irrational onus still exists. Currently, there is no objective evidence that masturbation is in any way physically harmful. Psychologically, it can serve useful purposes and is not considered pathological unless carried to extremes or used as a device to avoid social interaction. Masturbation can cause problems for some people from a moral standpoint because the practice is condemned by certain religious groups, especially Roman Catholics and Orthodox Jews. This proscription has not eliminated the practice within these groups, since youths associated with these particular religions do masturbate but less frequently than their peers in other groups. Indulgence under these circumstances can lead to anxiety and guilt feelings.

Heterosexual Activity

The primary function of the reproductive system is to produce individuals for the next generation. The first step in this process is the deposition of sperm into the vagina, normally during vaginal intercourse. That is all that is required as far as reproduction is concerned. There is no need for emotional contact between the two individuals or even physical contact (as in artificial insemination). Human sexual intercourse, however, involves much more than insemination. It can be a powerful emotional experience, a sensuous delight, and a learned skill. Of all the various synonyms for intercourse, the most expressive is "making love."

Most people learn the pleasures of heterosexual activity gradually. In an individual sexual encounter, the developmental sequence of holding hands, kissing, and petting (stimulation of the genitals and other erogenous zones) can be termed foreplay. Foreplay is physiologically desirable because it facilitates intercourse by lubricating the vagina and relaxing vaginal musculature. It is also an enjoyable prelude to this most intimate of human interactions. The subjects of foreplay techniques and intercourse are so extensive that a few simple comments cannot do justice to them. These topics are covered very thoroughly in many excellent manuals currently available. One should be aware, however, that these manuals concentrate heavily on technique rather than on the emotional phases of sexual activity and that loving and caring can transform a mechanical act into a considerably different experience.

The relationship of sexual partners is also a matter in which there is more freedom than in previous generations. Extramarital sex and live-in relationships are socially more acceptable now and consequently more common. Marriage is still the most popular basis for a long-term sexual relationship, but the 1980 census has shown that 1 in 30 couples in the United States are cohabiting, three times as many as in 1970.

Homosexuality

A person who prefers sexual activity with a partner of his or her own sex is **homosexual.** Commonly used synonyms are gay, homophile, and for women, lesbian. Although sexual orientation defines homosexuality, an undue emphasis on sex is as unfair to homosexuals as suggesting that sexual activity is the only feature of life that matters for heterosexual persons. Homosexuality is best considered a lifestyle in which intimate relationships are formed, by preference, with individuals of the same sex. The lifestyles vary considerably; some homosexual people prefer to live alone, whereas others have long-term, live-in relationships. Some may even be partners in heterosexual marriages, have children, but also have homosexual liaisons; that is, they are **bisexual.**

Homosexuality was accepted and even idealized by the Ancient Greeks and Romans but was severely condemned throughout most of the Judeo-Christian world. Even quite recently homosexuality was considered to be a pathological condition. Individuals with such inclinations were felt to need psychiatric attention in order to be "cured." They were also subject to social and legal harassment. Consequently, homosexual persons tended to be secretive, and the phenomenon of homosexuality was poorly understood. It is now considered a variation from the "norm" of heterosexual behavior and is more widely accepted as a legitimate lifestyle.

How and why does a person develop a homosexual preference? One commonly held theory is that male gays are products of a family life in which the mother is a strong, overprotective, even seductive person and the father is a relatively weak figure. This may explain some cases of homosexuality. It is also apparent that such family situations do not inevitably lead to the development of homosexuality. Another possibility is that the individual has simply learned, for various reasons, a sexual pleasure pattern with individuals of the same sex rather than with the opposite sex. Both of these concepts suggest that a homosexual preference is a learned pattern of behavior. There are some theories of homosexuality based on biological factors such as genetic tendencies, an endocrine condition, or some type of brain sex factor, though at present there is no solid evidence to support them. The problems with chromosomal, gonadal, or hormonal sex that have been discussed are, for the most part, serious enough to result in asexuality. There is thus no comprehensive, satisfactory explanation for the development of homosexuality.

Toward Sexual Fulfillment

Sexual feelings are a natural part of being human and a consequence of our biology and psychology; they are strongly affected by the familial, social, and religious influences that have shaped our lives up to now. Sexual feelings are neither urges that simply need periodic relief nor all-consuming forces that demand total dedication. Sexuality is a significant part of living that is best when integrated into the whole. The development and fulfillment of our sexual potential are as important as the growth of other capacities. Book learning can be helpful, but overintellectualization can be a handicap. The development of sexuality involves the recognition of some important emotions as well as of biological functions, and emotional growth comes primarily from personal experience.

CHAPTER 20

Human Reproductive Biology in the Future

Looking into a scientific crystal ball is a popular pastime, especially in the field of reproduction. I wish to review in this chapter a few developments that are actively being researched and may possibly be sufficiently perfected for use with humans during the next decade or two. Such new developments usually carry with them not only wonderful possibilities and opportunities but also the potential for abuse and the need for decisions and moral judgments by the individuals concerned and by society as a whole. There is often little precedent for such decisions, since the choices involved are different from those our parents faced. An important part of the preparation for these eventualities is accurate information.

Genetic Engineering

The discovery of the chemical makeup of genes immediately made the controlled alteration of human genes at least a theoretical possibility. The recombinant DNA techniques, a significant breakthrough of the 1970s, made the alteration of the chromosomes and genes of viruses and bacteria feasible and the creation of new strains of microorganisms possible. Genetic engineers are now able to transfer genes from one organism to another as well as to synthesize new genes and splice them into the DNA of living cells of higher animals. These experiments created considerable controversy at first, not primarily because of an ethical concern about the problems that the manipulation of human chromosomes might create but rather because of the fear that a new and very dangerous strain of bacteria or virus might be created and accidentally released into the environment. These fears have not been realized, and the genetic revolution continues to gather momentum. It is reasonable to expect that sometime in the near future the "repair" of human genes will be possible for the correction of serious single-gene defects such as hemophilia or sickle cell anemia.

Control of Sex of Offspring

This dream, which humans have had since prehistoric times, is close to being a reality. There are two types of sperm: those bearing a relatively large X chromosome and those with a much smaller Y chromosome. The sex of a child depends on which of these fertilizes the egg (Chapter 3). It should be feasible to control sex determination by (1) mechanically or chemically separating the two kinds of sperm in semen; (2) differentially stimulating one of these two types to greater activity; or (3) inhibiting, immobilizing, or filtering out one of these types. Poultry and beef farmers have a very practical interest in being able to do this and thus are supporting research in this area. At present there is a widely circulated formula for selectively stimulating sperm. It is based on the belief that X-bearing sperm are stimulated to swim better and faster in an acidic environment. Therefore, douching the vagina with a dilute vinegar solution prior to intercourse should increase the chance of conceiving a female child. Conversely, since Y-bearing sperm are supposed to swim better in a basic environment, a baking soda douche should increase the chance of conceiving a male child. The effectiveness of this method has not been substantiated, but undoubtedly more sophisticated and effective methods of accomplishing the same results will be made available within the next decade. There are some who are concerned about the implications of altering the sex ratio in a society, but it is doubtful that their expressed fears will have much impact on this research.

Writing about this type of control over reproduction in the future tense is somewhat misleading, since a foolproof method of guaranteeing the sex of an offspring is currently available. It is, however, relatively drastic and undoubtedly morally offensive to most people. The sex of an unborn child can be accurately determined by a chromosomal analysis of cells obtained by amniocentesis (Chapter 16). An abortion can then be performed if the sex is not that which is desired. Some obstetricians have emphatically stated that they would never perform such tests for that purpose. Nevertheless, the technology now exists, and a determined woman probably could manage to avail herself of it.

Test-Tube Babies, Surrogate Motherhood, and Cloning

Sexual activity has inevitably led to reproduction for millenia of human history. Current advances in contraceptive technology as well as the free dissemination of information about contraception—a very recent phenomenon—have led to a certain degree of dissociation between sexual activity and reproduction. Almost all children of future generations will continue to be conceived in the same fashion as their ancestors, but some will not. Already, children conceived by artificial insemination, whether by husband or by donor, or by the test-tube baby technique mark the beginning of a new era in which reproduction can be entirely separated from sexual activity.

Successful *in vitro* fertilization of human eggs and test-tube babies are now practical realities (Chapter 18). These techniques can eventually be used in additional ways. Animal experimentation has demonstrated that implanting a cultured embryo into a female other than the ovum donor—a **surrogate mother**—works very effectively. There is no reason to expect that there are any biological impediments to surrogate motherhood in humans. A woman who wanted her own genetic offspring but could not carry a baby because of a uterine, cardiac, or kidney problem, or who simply did not want to bother with a pregnancy, could avail herself of this technique. All that would be required is that (1) she has adequately functioning ovaries and (2) she can find a volunteer to be a surrogate mother. The surrogate would carry the baby for nine months and then turn over the child to the natural parents after birth (Fig.

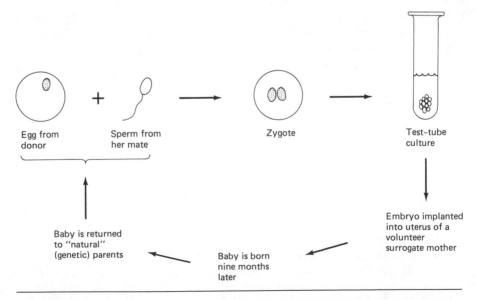

Figure 20–1   A schematic representation of the surrogate motherhood procedure.

20–1). Although such "interference with nature" might concern some individuals, it is the potential use of the test-tube baby technique, discussed next, that creates most of the concern about this technology.

A **clone** is a group of organisms with identical genetic constitutions derived from a single individual. Since sexual reproduction results in a mixing of the genetic contributions of both parents, a clone must be obtained by using some type of asexual reproduction (i.e., reproduction by methods other than sexual ones). A group of plants propagated by cuttings from a single mother plant is an example of a clone. Identical twins are a clone of two because they came from a single egg; an armadillo litter is a clone of four for the same reason. Cloning has been done successfully in the laboratory with frogs, and uniparental mice have been created by a technique that might someday yield a clone of mammals. The frog technique is described in the following paragraph because a modification of it, along with the surrogate motherhood technique, may make human clones possible in the future.

Frog clones are made by removing the nucleus of an egg and replacing it with a nucleus from a mature cell of a donor (Fig. 20–2). The eggs are first

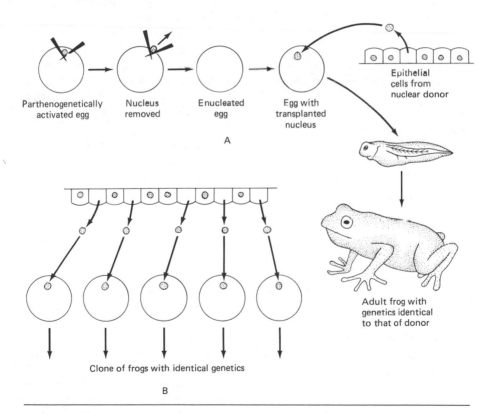

Parthenogenetically
activated egg

Nucleus
removed

Enucleated
egg

Egg with
transplanted
nucleus

Epithelial
cells from
nuclear donor

A

Adult frog with
genetics identical
to that of donor

Clone of frogs with identical genetics

B

Figure 20–2 **Nuclear transplantation in frogs. *A*, The experimental technique. *B*, Obtaining a frog clone.**

activated with an electric shock, which can mimic the effect of sperm contact. This brings the female nucleus to the surface of the egg, where it can be removed with a pair of needles (Fig. 20–2A). This egg, though now ready to develop, cannot do so because it has no nucleus. A nucleus is removed from some mature tissue, such as a layer of epithelial cells, and is injected into the enucleated egg. The egg then develops into a tadpole and eventually into an adult frog. This frog has no father because no sperm were used. It also is not genetically related to the female that laid the egg because that nucleus was removed. It is, however, a precise genetic copy, gene for gene, of the frog that provided the donor nucleus. To produce a clone, a group of animals with identical genetics, this procedure is simply repeated with 5, 20, or even 100 enucleated eggs into which mature nuclei from the same donor are transplanted (Fig. 20–2B). This experiment has been done numerous times with frogs.

The attempts to perform direct nuclear transplantation in other animals have not been successful, despite considerable effort. There are, however, other possible ways in which the same net result could be accomplished, that is, the replacement of the female pronucleus of an egg with a mature diploid nucleus from another individual. One method that has been proposed is to use a virus of the Sendai strain, which affects cell membranes in such a manner that they fuse together like sperm and egg. A nucleated cell, obtained, for example, from a scraping of the inside of the cheek or from a drop of blood, could be fused in this manner with an enucleated egg and thus introduce an adult nucleus into the egg (Fig. 20–3). This method works well with various mature cells, but it has not yet been done successfully with a mammalian egg. It is still a possibility.

The potential misuse of human cloning, if it is ever perfected, has genuinely frightened a number of serious-thinking individuals. The specter of the creation of a clone of Frankenstein monsters, or of a master race, is not countered by the possibility of creating a clone of Einsteins. All experiments with human eggs have been condemned on a moral basis by several religious and professional groups. Such a blanket censure has no chance of being effective. The experiments with human eggs will continue because they have a laudable goal—a greater understanding of human reproduction and the possibility of helping some infertile couples to have children of their own. The experiments with nuclear transplantation and cell fusion will proceed because they are a vital tool in understanding cellular genetics and such major problems of cellular biology as cancer. Even the development of a human clone could help to resolve with some degree of finality the intriguing and perennially recurrent question of the relative importance of heredity and environment in determining various parameters of intelligence and behavior. It must be recognized that (1) the basic technology for human cloning already exists and will continue to be developed because it is valuable for many aspects of basic and applied medical research, (2) this possibility is so intriguing that it will be attempted, and (3) sooner or later a success will occur. Therefore, there are only two questions

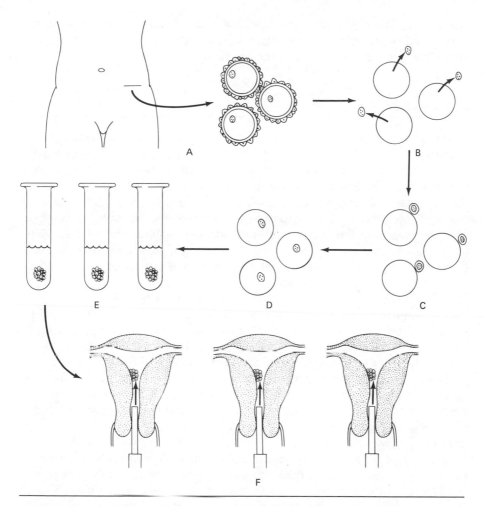

Figure 20–3 Potential method for obtaining a clone of humans. *A*, A number of eggs are surgically obtained from one or more donors. *B*, These eggs are then stripped of corona cells, activated with an electrical shock, and enucleated. *C*, The enucleated eggs and a group of cells from a single donor are treated with Sendai virus. *D*, Fusion of eggs with normal cells results in "fertilized" eggs with regular (diploid) nuclei from cell donor in them. *E*, The eggs are cultured for several days to the morula stage. *F*, The embryos are then transplanted into the uteri of a series of surrogate mothers.

A clone of identical individuals could be born nine months later. These babies would be totally unrelated to the women who supplied the eggs or to the mothers who bore them and would have no biological fathers. But they would be absolutely genetically identical to the individual who supplied the donor cells with nuclei. Steps *C* and *D* are the only ones that have not yet been done successfully with mammalian eggs, though they have been performed with other cells.

that remain: What limitations does society wish to place on this type of research and, even more important, how can such limitations be enforced? There are no simple answers to these questions.

Epilogue

In writing this book, I have tried to treat the various aspects of human reproduction with scientific objectivity. I am well aware that phases of human activity such as intercourse, pregnancy, birth, reproductive problems, and birth defects can have considerable emotional impact on individuals, but a detailed discussion of these psychological considerations is beyond the immediate scope of this book. My primary aim was to write a comprehensive text about the biology of human reproduction and development that can be understood by individuals with a limited scientific background. Another goal was to provide my readers with a solid factual basis for making sound decisions about any reproductive crisis that may be encountered.

I have not been unmindful, either, of the ethical problems that are created by amniocentesis, test-tube babies, cloning, and other new developments in reproductive biology. It is axiomatic that scientific technology progresses at a rate that is considerably faster than society's ability to cope with the problems that result from these innovations. For example, many of the ethical and legal problems of artificial insemination, a relatively simple technique practiced for decades, have not yet been solved. Not all technological advances are unmixed blessings. Almost any scientific advance that is helpful for some individuals will be accepted into our technological repertoire, along with the sociological and ethical problems that it provokes. Many recent advances have alleviated problems that have vexed individuals and groups for generations, but they have also raised perplexing ethical questions. For instance, heterozygous carriers of many serious genetic diseases can now be identified. Should these individuals have offspring or should they voluntarily remain childless? If they decide to have children, can they agree to abort any homozygous recessive child that might be prenatally detected? Even if they do, is it fair for them to pass these genes on to their symptomless heterozygous children, along with the responsibilities that come with the knowledge of these circumstances? There are no easy answers to these and other ethical questions posed by new technology. The best preparation for these new questions, for which traditional answers are not applicable, is a solid body of factual information. The most satisfying answers will come from informed individuals.

To the readers who pursue this information, this book is affectionately dedicated.

Glossary of Technical Terms

A

Abortion A pregnancy terminated before the fifth month of development either by spontaneous factors or by deliberate intervention.

Acrosome A specialized region in the head of a sperm containing enzymes that digest the membranes around the egg and permit sperm penetration.

Activation The release of the dormant biochemical mechanisms for embryonic development stored within the egg, usually initiated by the sperm.

Adrenal gland An endocrine gland located above the kidney that secretes several hormones, including small quantities of estrogens and androgens, in both sexes.

Amniocentesis The withdrawal of amniotic fluid from a pregnant woman for the purpose of chemical or chromosomal analysis.

Amnion The fluid-filled membrane that surrounds the embryo and fetus, providing mechanical protection and a space in which to develop. It is also known as the "bag of water" that ruptures just before birth.

Androgens The generic name for a group of steroid compounds, both natural and synthetic, that have male hormone activity.

Andrology The study and treatment of male fertility problems.

Aorta The large artery that comes directly from the left side of the heart and delivers blood to the body.

Aortic arches A series of six pairs of blood vessels in the head and neck area of the embryo that eventually develop into the major arteries arising from the heart, that is, the aorta, carotid, and subclavian arteries.

Artery A blood vessel that delivers blood from the heart to an organ.

Artificial insemination The placement of an ejaculate into or around a woman's cervix with a syringe or other device. The semen can be obtained from the husband (AIH) or from an anonymous donor (AID). It is used in cases of male fertility problems.

233

Atria (singular: atrium, an entrance) The two chambers of the heart that receive blood from either the body or the lungs and deliver it to the ventricles.

Autonomic nervous system That part of the nervous system that controls internal (visceral) functions.

Autosexual activity Sexual activity performed without a partner, such as fantasy and masturbation.

Autosomes The 22 pairs of chromosomes in a human cell that are not sex chromosomes.

B

Bartholin's glands A pair of glands located adjacent to the vaginal orifice. They secrete a fluid that keeps the vestibule moist.

Basal body temperature (BBT) A method for detecting ovulation time either to avoid or to enhance the possibility of pregnancy. It is based on the fact that body temperature rises slightly, but consistently, at ovulation.

Biopsy Removal of a slice of tissue from the body for the purpose of diagnosis by microscopic examination. Frequently used to determine if a tumor is benign or malignant.

Birth defect An anatomical or functional abnormality that results from a developmental problem and may have been induced by genetic or environmental factors.

Bisexual An individual who indulges in both heterosexual and homosexual activity.

Blastocyst The hollow ball stage of human development, which occurs about a week after fertilization.

Bonding The establishment and affirmation of psychological ties between mother and child that occurs from physical contact and nursing.

C

Cancer See Tumor.

Candidiasis A vaginitis induced by a fungus infection; it is also called moniliasis.

Capacitation The final changes that occur on the sperm head over a period of several hours in the female reproductive tract. These must take place before the sperm can fertilize the egg.

Carotid arteries Blood vessels that extend from the arch of the aorta to the head and neck, delivering blood to these regions.

Carrier An individual who, though symptomless, can transmit a disease, either hereditary or infectious.

Castration Removal of the testicles.

Cervical Adjective referring to parts of the neck, for example, cervical vertebrae. It also can refer to the cervix (neck) of the uterus.

Cervical cap A contraceptive device, a thimble-shaped cup that fits over the cervix.

Cervical dilation The process of distention of the cervical canal, which occurs spontaneously but slowly before the delivery of a baby (first stage of labor) or which may be done artificially before an induced abortion or uterine surgery such as D & C (dilatation and curettage).

Cervix The neck of the uterus, which extends into the deep portion of the vagina. It is traversed by a cervical canal, which is the opening into the body of the uterus.

Cesarean section The surgical removal of a baby through incisions in the abdominal wall and uterus.

Chancre The hard, round ulcer that appears at the site of a syphilitic infection.

Chromosomes The organelles within a nucleus that contain the genes. Each human cell contains 46 chromosomes arranged as 22 pairs plus a special pair of sex chromosomes.

Cilia The microscopic hairlike extensions of epithelial cells, which beat in wavelike rhythms and are thus capable of moving small objects in internal organs.

Cleavage The period immediately following fertilization during which the egg divides very rapidly without growing in size or changing shape.

Cleft lip A defect in facial development that results in a gap in the upper lip to either side of the midline area (philtrum).

Cleft palate A gap in the roof of the mouth due to the failure of the palatine bars to fuse during development.

Clitoris An external genital organ in the woman that contains erectile tissue and many sensory nerve endings; it is the chief site of sexual sensation in the woman, homologous to the penis of the man.

Clone A group of individuals with identical genetic constitutions.

Coitus The act of sexual intercourse.

Colostrum A clear liquid secreted by the breasts right after delivery, before mature milk is produced a few days later.

Condom A contraceptive device; a thin sheath, usually rubber, that envelops the penis during intercourse, thus preventing the semen from entering the vagina.

Conjoint twins Twins that are born attached to each other to some extent; they are also known as Siamese twins.

Connective tissue One of the four basic tissues, specialized to interconnect various other cells and tissues, typically consists of cells and interlacing fibers.

Contraception A temporary method for preventing pregnancy.

Contraceptive effectiveness A measure of the ability of a contraceptive to prevent pregnancy, expressed as failure rate (i.e., pregnancies) per 100 woman years of exposure. Theoretical effectiveness is a measure of the optimum success of a method. The use effectiveness is the success rate of a method as used by an average cross section of a population.

Corona radiata   The crown of follicle cells continuing to surround the egg after it is released from the ovary.

Corpora cavernosa   The spongy tubes that extend from the shaft of the penis to their internal anchorages. They become turgid and produce an erection when they fill with blood.

Corpus luteum   Latin for "yellow body"; this is the organelle that develops from follicle cells after the egg is released. It synthesizes progesterone as well as estrogen.

Cortex   The outer layer (bark) of many organs, such as the ovary, in distinction to the medulla, or center.

Cortisone   A hormone secreted by the adrenal cortex.

Cremaster   A sheet of muscle that extends from the abdominal wall and surrounds the testes. An important element in the temperature-regulating mechanism of the testes.

Critical period   A time at which an organ is particularly sensitive to insults, usually early in development.

Cryptorchidism   A condition in which a testicle fails to complete its descent into the scrotum, usually becoming lodged in the inguinal canal.

Cystitis   An infection of the urinary bladder.

Cysts   Sacs in some abnormal location that are filled with a fluid or semi-solid substance.

Cytology test   A microscopic examination of cells obtained from an organ to test for cancer or precancer stages; it is also called Pap smear.

Cytoplasm   A major component of a cell, the peripheral portion where the actual metabolic work of a cell is performed.

D

Diabetes   A disease caused by a deficiency of insulin.

Diaphragm   A contraceptive device, a disc of thin rubber that loosely covers the cervix. It is covered with spermicidal jelly and placed in the vagina prior to intercourse.

Diethylstilbestrol (DES)   A synthetic estrogen.

Dilatation and curettage (D & C)   A gradual dilation of the cervix followed by a scraping out of the contents of the uterus with a tool called a curette. Used for first trimester abortions and for some menstrual disorders. See also *Cervical dilation*.

Diuresis   Copious urination.

Dizygotic twins   Twins that have developed from two separately fertilized eggs and hence are no more closely related to each other than a brother and sister might be; also known as fraternal twins.

DNA   Deoxyribonucleic acid. The chemical name of the large molecular chain (double helix) of which genes are composed.

Dominant characteristic   One that is controlled by a dominant gene, the trait that appears in a person who is homozygous or heterozygous for that characteristic.

Dominant gene   The member of a gene pair that is more potent in exerting an effect, to the degree of totally masking the recessive gene, should it be present.

Dorsal   The back side of the body.

Douche   A current of water projected into a cavity—for example, vaginal douches.

Down's syndrome   A severe form of mental retardation caused by a chromosomal abnormality (trisomy of Number 21).

Ductus arteriosus   The blood vessel in the fetus that interconnects the pulmonary trunk and aorta, bypassing the fetal lungs.

E

Ectoderm   The most superficial of the three basic embryonic tissues; it develops into the skin and nervous system.

Ectopic pregnancy   A condition in which a blastocyst attaches to the oviducts or some part of the body other than the body of the uterus and continues to develop there.

Edema   An excessive accumulation of extracellular fluid that causes swelling of tissues.

Egg   The female reproductive cell, the product of the ovary that has the capacity to become fertilized and develop into a new individual.

Ejaculation   The discharge of semen during a male orgasm.

Embryology   The field of biology that studies the development of individuals.

Endocrine glands   Glands that synthesize hormones and deliver them directly into the blood stream. They form a system of chemical coordinators of bodily processes.

Endoderm   The deepest of the three basic tissues of the embryo; it develops into the linings of the lungs, intestine, and other internal organs.

Endometrium   The epithelial lining of the lumen of the uterus.

Enzymes   Proteins that catalyze (promote) chemical reactions in the body.

Epididymis   A part of the sperm transport system; it is a network of tubules that forms a cap to one side of the testicle and is connected with the rete testis on one end and the vas deferens on the other.

Epigenesis   New formation; this is the process of development by which a body forms from formless precedents, grows from a simple to complex entity.

Episiotomy   A surgical incision from the posterior edge of the vagina and into the perineum, which is often made just before the delivery of the baby's head to prevent the accidental tearing of vaginal tissue.

Epithelium   One of the four basic tissues, usually arranged as a continuous layer for the purpose of lining or covering some part of the body.

Erythroblastosis fetalis A disease of newborns caused by Rh incompatibility.

Estrogens The generic name for a group of steroid compounds, natural or synthetic, that have female hormone activities, such as maintenance of the reproductive organs and production of secondary sex characteristics.

Eunuch A male who has been castrated before puberty.

Evagination An outpocketing of cells, a form-building process by which an organ can originate from a flat sheet or tube of cells.

Experimental embryology The field of study that emphasizes analysis of the causes of development as determined by experimental methods.

Extraembryonic membranes Cellular extensions of the embryo that support its development, for example, amnion and yolk sac.

F

Fallopian tubes *See Oviducts.*

Fascia A tough sheet of connective tissue fibers that helps to bind organs together.

Feedback control A check-and-balance system that helps to maintain the blood levels of several hormones within a narrow range.

Fertilization The sequence of events that starts when the sperm contacts the egg and ends when the nuclear contributions of egg and sperm fuse together.

Fetal alcohol syndrome (FAS) A combination of mental retardation, growth deficiency, and facial abnormalities caused by prenatal exposure to alcohol.

Fetus An individual between the third and ninth month of prenatal development in which most organ formation has already taken place and growth and maturation are the principal processes occurring.

Fibroid A benign tumor of the myometrium.

Folding Embryonic form-building process by which an organ—for example, the brain—can develop from a flat sheet of cells.

Follicle The ovarian organelle within which the egg finishes its last period of growth and maturation.

Follicle-stimulating hormone The pituitary gonadotrophin that stimulates follicle formation in the woman and sperm production in the man.

Foramen An opening.

Foramen ovale The oval opening between the two atria, which, with a thin, perforated membrane (the first interatrial partition), constitutes the interatrial flutter valve.

Foreskin *See Prepuce.*

Fraternal twins *See Dizygotic twins.*

Frigidity Sexual unresponsiveness of the woman.

Frontal eminence The front part of the embryonic face that gives rise to the forehead and nose.

G

Gametes   The reproductive cells; a collective term for both sperm and egg.

Gender identity   The personal awareness of being a female or male individual.

Gender role   The public expression of one's concept of her or his gender identity.

Genes   The units of heredity that are located on the chromosomes of the nucleus of cells.

Genetic engineering   Altering the genes of a chromosome by technical means, a technique currently used on bacteria and viruses.

Genetics   The biological science that is concerned with the transmission of hereditary characteristics.

Genital swellings   A pair of mounds in the pubic area that develop into the scrotal sacs in the male individual and the labia majora in the female individual.

Genital tubercle   The embryonic phallus, which is equally prominent in both sexes, develops into the penis in the male individual and the clitoris in the female individual.

German measles   See *Rubella*.

Germinal disc   In a two-week human embryo, the region between the amnion and yolk sac, which develops into the body of the embryo.

Gland   A group of cells, usually of epithelial origin, specialized to secrete a substance.

Glans penis   The head of the penis, somewhat larger in diameter than the shaft of the penis.

Gonadotrophin   A pituitary hormone that affects the gonads. There are two of these: luteinizing hormone (LH) and follicle-stimulating hormone (FSH). LH is also known as interstitial cell–stimulating hormone (ICSH).

Gonorrhea   A venereal disease induced by the bacterium *Neisseria gonorrhoeae*.

Gynecology   The branch of medicine that deals with problems of the female reproductive organs.

H

Harelip   See *Cleft lip*.

Hemophilia   A genetic disease in which there is a deficiency of the blood-clotting mechanism; consequently, even small injuries can produce excessive bleeding.

Hermaphrodite   An individual who has both ovarian and testicular tissue; it is an extremely rare occurrence in humans. A pseudohermaphroditic condition is more common.

Hernia   An abnormal protrusion of abdominal contents into a region; it is also called a rupture. The most common site of hernias in men is the inguinal canal and scrotum.

Herpes genitalis A virus infection (herpes simplex, type II) that causes small blisters and ulcers over the genital area and is a sexually transmitted disease (STD).

Heterosexual An individual who prefers a person of the opposite sex for a sexual partner; the term also pertains to sexual activity between persons of opposite sex.

Heterozygous Differing with respect to a given pair of genes, for example, *Hh* and *Bb*.

Homeostasis A physiological mechanism by which some bodily function such as temperature, blood sugar, and hormone level is maintained at a steady level.

Homologous Refers to structures that have the same origins.

Homosexual An individual who prefers a person of the same sex for a sexual partner; the term also pertains to sexual activity between persons of the same sex.

Homozygous Alike with respect to a particular gene, for example, *HH* and *bb*.

Homunculus The preformed individual that supposedly was contained in the egg or sperm according to the seventeenth-century concept of preformation.

Hormones Chemicals that are synthesized by endocrine (ductless) glands, are delivered to the blood stream, and produce significant effects on various parts of the body.

Human chorionic gonadotrophin Abbreviated hCG, this is a hormone of pregnancy that is produced before implantation and prevents the degeneration of the corpus luteum. Later it is synthesized by the placenta and stimulates estrogen and progesterone production in the ovary. It is also the hormone that is detected in pregnancy tests.

Human placental lactogen Abbreviated hPL, this is the hormone produced by the placenta that stimulates the growth of the breasts and prepares them for lactation.

Hydrocele An abnormal collection of fluid in the scrotum.

Hymen A membrane that partially occludes the vaginal orifice in young girls.

Hypothalamus A small region of the brain, just above the pituitary, that secretes the releasing factors that control pituitary functions.

Hysterectomy Surgical removal of the uterus.

I

Identical twins See *Monozygotic twins*.

Implantation The process by which the embryo (blastocyst) becomes embedded into the uterus, occurring during the second week after fertilization.

Impotence Erectile dysfunction—the inability to achieve and maintain an erection during sexual activity.

Induction A type of cellular interaction by which one group of cells stimulates the initial development of a new organ in another group of cells of an embryo during a short period of contact between the two.

Inferior vena cava   The large vein that collects blood from the trunk of the body and returns it to the right atrium.

Inguinal canal   An oblique passageway through the abdominal musculature just above the pubis through which the testicles migrate before birth.

Inner cell mass   The group of cells within a blastocyst that eventually develops into the embryo.

Inner ear   The innermost part of the ear embedded within the skull, which consists of the cochlea, semicircular canals, and other parts.

Insult   As used in this context, a physiological shock to an embryo or fetus that can produce birth defects or other types of damage.

Interatrial flutter valve   The combination of the foramen ovale and the perforated first interatrial partition, which allows blood to move from the right atrium to the left atrium, but not in the other direction. One of the two systems by which blood in the fetus can bypass the lungs.

Interstitial cells   The large cells located between the seminiferous tubules of the testes. These are the cells that secrete the male hormone.

Interstitial cell—stimulating hormone   See *Luteinizing hormone.*

Interventricular septum   The partition that develops between the two ventricles of the heart.

Intrauterine device (IUD)   A contraceptive device consisting of a loop, disc, or coil of flexible metal or plastic that is inserted into the uterus, where it prevents implantation of the blastocyst.

Invagination   An inpocketing of cells, a form-building process by which an organ can originate from a flat sheet or tube of cells.

Ischemia   A lack of oxygen to some tissue or organ.

—itis   A suffix that indicates inflammation. When added to the name of an organ (e.g., vaginitis or prostatitis), it indicates inflammation of that organ from infection, mechanical trauma, or other causes.

L

Labia majora   A pair of fatty folds of flesh that enclose all the other external genitalia of women.

Labia minora   A pair of fleshy folds just inside the labia majora that enclose the vestibule.

Lactation   The period after pregnancy during which the breasts secrete milk.

Laryngotracheal groove   The ventral evagination of the primitive gut, which gives rise to the larynx, trachea, and lungs.

Larynx   The voice box, which is located in the neck.

Libido   The sexual drive, the desire for sexual activity.

Limb buds   The microscopic elevations of tissue on the flanks of the embryo that develop into arms and legs.

Locus   A point, for example, a specific, small region on a chromosome that is the site of a single gene.

Lumen   The central open portion of a hollow organ.

Luteinizing hormone   The gonadotrophic hormone that controls progester-one synthesis in the woman and testosterone synthesis in the man. Also known as interstitial cell–stimulating hormone (ICSH).

M

Mammary glands   The glandular (secreting) tissue of the breasts.

Mandibular arch   The greater portion of the first visceral arch, which develops into the mandible (lower jaw).

Masturbation   The physical stimulation, usually by oneself, of the genitalia leading to orgasm.

Maxillary arch (segments)   The wedges that separate from the first visceral arch and give rise to the cheeks and lateral parts of the upper jaw.

Medulla   The central portion of an organ, in distinction to the cortex, or outer layer.

Meiosis   A special form of cell division limited to the gametes and character-ized by a reduction of chromosome content by one half.

Menarche   That time during female puberty at which menstruation begins.

Menopause   That period in the life of a woman when estrogen secretion ceases and reproductive senescence occurs.

Menstrual extraction   Removal of uterine contents just before menstruation by gentle suction, either to avoid the nuisance of a menstrual period or to remove a possible very early pregnancy.

Menstruation   The flow of cast-off endometrial layer and blood that occurs at about four-week intervals in women during their reproductive years; it is also known as menses.

Mesoderm   One of the three basic tissues of the embryo; this middle layer develops into muscle, bone, and many internal organs.

Mesonephric duct   The tube that drains the mesonephros of urine; it becomes the vas deferens in the male.

Mesonephros   An embryonic kidney that is functional throughout much of embryonic and fetal life and then disappears, except for some portions that become parts of the male reproductive system.

Messenger RNA   The chemical signal manufactured in the nucleus that then goes into the cytoplasm to control the metabolic activity there.

Metanephros   The kidney of adult humans.

Minipill   An oral contraceptive that contains only progesterone and no estro-gen.

Mitochondria   A subcellular organelle that contains the enzymes for the generation of energy for cellular functions.

Mitosis   The process by which ordinary cell division occurs. It is character-ized by a cytoplasmic division and a single duplication and separation of chromosomes, so that the net result is the formation of two daughter cells, fully equal with respect to cytoplasm and the genetic content of the nucleus.

Monozygotic twins Twins that have developed from a single fertilized egg (zygote) and hence have identical genetic constitutions, sex, and so on; they are also known as identical twins.

Mons veneris A pad of fatty tissue above the labia major that covers the pubic bone.

Mucous membrane See *Mucus*.

Mucus A clear, viscous secretion, commonly secreted by glands to keep internal surfaces moist. An epithelial layer kept moist in this way is a mucous membrane.

Müllerian ducts Paired tubes of the embryo that develop into the oviducts and uterus normal of the female individual and disappear in the male individual.

Muscle One of the four basic tissues consisting of cells specialized for contraction.

Myometrium The layer of smooth muscle in the uterus.

N

Nasal processes The two ridges on either side of the embryonic nasal pits.

Natal Relating to birth, often used as a suffix—for example, perinatal, meaning "around birth," and postnatal, meaning "after birth."

Neonate A newborn individual.

Nerve cells See *Neurons*.

Neural tube The simple tube of tissue in the embryo that develops into the brain and spinal cord.

Neurons Nerve cells; these cells are specialized to conduct impulses.

Notochord A rodlike embryonic organ that lies underneath the early neural tube.

Nucleus A major component of a cell, the central portion; it contains genetic material and is the region where chemical signals from genes are synthesized and then relayed to the cytoplasm, where they control cellular activity.

O

Oral contraceptive Literally, any contraceptive that is taken by mouth; but as used currently, this phrase, along with the term "the Pill," refers to preparations of synthetic estrogens or progesterone or both that inhibit ovulation and hence prevent pregnancy.

Organ An anatomical unit of the body having a definite form and one or more specific functions.

Organizer center A region in the embryo that exerts great influence on the development of surrounding areas, for example, the primitive node.

Orgasm The collective sensations that occur at the height of intercourse, marked by ejaculation in the man and, in both sexes, by rhythmical contractions of pelvic musculature.

Orifice An opening of or entrance to an organ.

Ovary The female sex gland; it produces egg cells and female reproductive hormones and lies within the pelvic cavity.

Oviducts The tubular structures that interconnect the ovary and uterus for the purpose of transporting egg and sperm.

Ovulation The process by which the follicle ruptures, releasing the egg and follicular fluids.

Ovum (plural: ova) The cellular product of the ovary; it is the only cell of the body that can develop into a new individual.

Oxytocin A hormone of the posterior pituitary gland that stimulates uterine muscle to contract and milk to be ejected from the breasts.

P

Palate The roof of the mouth, which separates the oral and nasal cavities.

Palpation Examination of an organ by feeling with the fingertips.

Pap smear See *Cytology test.*

Patho— A prefix that refers to disease; for example, *pathology* means study of disease; *pathogenic* means disease-producing.

Pathogenic Disease-producing.

Pelvic inflammatory disease (PID) An infection that ascends the female reproductive tract, through the vagina, uterus, and oviducts, and extends into the internal pelvic organs.

Pelvis The large, funnel-shaped bone at the base of the spinal column.

Penis The male copulatory organ.

Perineal body A small region of the external body between the anus and external genitalia.

Pharynx The portion of the alimentary canal that is in the head, extending from the back of the nose and throat to the larynx. Also, the embryonic portion of the primitive gut that gives rise to these regions.

Philtrum The center area of the upper lip.

"Pill" See *Oral contraceptive.*

Pituitary gland A small endocrine gland located at the base of the brain that produces a large number of hormones.

Placenta The disc-shaped region in the pregnant uterus where fetal and maternal blood streams are in juxtaposition and the interchange of oxygen, nutrients, and other vital substances occurs.

Polar bodies The small units of cytoplasm and nuclear material that are cast off during meiosis of eggs.

Polyps Tumorous outgrowths of mucous membranes, characteristically attached to the membrane by a narrow stalk.

Preeclampsia A serious problem of late pregnancy that is characterized by fluid retention and high blood pressure. If untreated at early stages, it can lead to convulsions and other problems.

Preformation The seventeenth-century theory that the egg (or sperm) contained a preformed miniature individual and that development consisted simply of an expansion in size.

Prepuce   The loose skin that covers the glans penis. This is the skin that is removed in circumcision. It is also called foreskin.

Primary sex cords   Strands of tissue in the primitive bipotential gonad that develop into the seminiferous tubules of male individuals and disappear in female individuals.

Primitive gut   A thin tube of endodermal cells that develops into the intestinal tract and many other internal organs.

Primitive node   The enlarged portion at the anterior end of the primitive streak.

Primitive streak   A strand of tissue in the germinal disc stage of development (about one and a half to two weeks after fertilization) that eventually disappears after playing an important role in the formation of the primitive body axis.

Primordial germ cells   The precursors that form the stem cells, which in turn form sperm or eggs. They appear in the embryo before the gonads are present, circulate through the embryo, and migrate into the gonads shortly after they develop.

Primordium   A small group of embryonic cells marking the beginning of the development of a new organ.

Progesterone   A steroid female hormone necessary for the maintenance of pregnancy and for control of the menstrual cycle.

Prolactin   A pituitary hormone that stimulates milk formation in the breast after delivery.

Pronuclei   The partial nuclei of either male or female origin present in the egg just before the time they fuse to form the full nucleus of the new individual.

Prostaglandins   Substances found in the male ejaculate that stimulate the contraction of uterine muscle. They are also used to induce abortion and are instrumental in several kinds of contraceptive methods.

Prostate   Gland in the man that surrounds the urethra just as it leaves the bladder. About the size and shape of a chestnut, it contributes up to 60 percent of the volume of the ejaculate.

Prostatitis   An inflammation of the prostate gland caused by infection, cancer, or other causes.

Protein   Large, complex molecules composed of amino acids and synthesized by cells.

Pseudohermaphrodite   An individual who has either testes or ovaries but whose external genitalia show some characteristics of both sexes.

Puberty   The period of life when the reproductive system undergoes its final maturation.

Pubic lice   Blood-sucking insects that may infest the genital area.

Pulmonary artery   The blood vessel that arises from the right side of the heart and delivers blood to the lungs.

R

Recapitulation A biological concept referring to the observation that the individual, during his or her own development, often repeats imperfectly some of the stages of evolution of the species.

Recessive characteristic A trait that requires homozygosity at that locus to be expressed; the trait does not become expressed in a heterozygous individual.

Recessive gene The gene that is not expressed (i.e., masked) when present with a dominant gene for a particular trait in a heterozygous individual.

Releasing factors Hypothalamic hormones that control the synthesis of gonadotrophins by the pituitary. There is a specific releasing factor for each pituitary hormone controlled.

Rete testis A network of tubules on one end of a testicle that receives the sperm from the seminiferous tubules and, in turn, deposits them in the epididymis.

Rhythm method A system of avoiding pregnancy by not having intercourse for three to five days before and after estimated ovulation time. This six- to ten-day period is the time of *abstinence*. The remainder of the cycle is called the *safe period* because the possibility of conception is appreciably reduced.

Rubella German measles; this is a virus disease that can induce a high frequency of birth defects.

S

Safe period See *Rhythm method*.

Scrotum In the perineal area of the man, a pouch of skin containing the testes.

Secondary sex characteristics The hormone-mediated differences between the sexes in hair pattern, fat distribution, voice, and so on.

Semen The mixture of sperm and fluids from the seminal vesicles and prostate gland, which is ejaculated during a male orgasm.

Seminal vesicles Glandular male organs located at the junction of the vas deferens and urethra. They secrete sugar, which acts as an energy source for sperm, prostaglandins, and other substances, which constitute 30 percent of the ejaculate.

Seminiferous tubules The coiled tubules within the testicles where sperm production occurs.

Septum A membrane that forms a partition within an organ.

Sex chromosomes The special pair of chromosomes responsible for sex determination. They form a matched pair (XX) in women and an unmatched pair in men (XY).

Sex-influenced trait A genetic characteristic whose expression is affected by the sex of the individual, for example, pattern baldness.

Sex-linked characteristics A genetic trait that is controlled by a gene on the X chromosome. Since women have two X chromosomes and men only one, the pattern of inheritance of these traits differs in the two sexes.

Sexual response   The physiological changes that occur in the external genitalia and pelvis during sexual stimulation and intercourse, including pelvic vasocongestion, penile erection and ejaculation, vaginal lubrication, and orgasm.

Sexuality   The sum total of biological, psychological, and sociological aspects of sex and sex-related activity.

Sexually bipotential stage   That period of embryonic development (about the fifth week) in which the rudiments of the reproductive organs of both sexes are present in each individual.

Sexually transmitted diseases (STD)   Diseases that can be transmitted by sexual contact. These include venereal diseases as well as other disorders that are sometimes transmitted in this fashion.

Siamese twins   See *Conjoint twins.*

Skeletal muscle   A form of muscle consisting of very large, multinucleated cells arranged into long fibers. This type of muscle contracts rapidly, is ordinarily under voluntary control, and is generally associated with the movement of the skeleton and other external parts of the body.

Skene's glands   A pair of glands in women located at either side of the urinary meatus that secrete a fluid that helps keep the area moist.

Smooth muscle   A form of muscle consisting of individual spindle-shaped cells. These muscles contract slowly and are not ordinarily under voluntary control. They are associated with the movement, often rhythmical, of internal organs.

Somites   Aggregations of mesodermal cells that develop into the vertebrae of the spinal column.

Sperm   The male reproductive cell, a product of the testis, which has the capacity to fertilize the egg.

Spermatic cord   A complex consisting of the vas deferens, artery, vein, and nerve, which extends from the inguinal canal to the testes.

Spermatogonia   The stem cells that develop into sperm; they are located around the periphery of a seminiferous tubule.

Spermicide   A chemical preparation that renders sperm incapable of fertilization; it may be in the form of cream, jelly, or foam and is usually placed in the vagina before intercourse.

Spirochete   The spiral-shaped microorganism that causes syphilis.

Spotting   Uterine bleeding that occurs at some point between menstrual flows.

STD   See *Sexually transmitted diseases (STD).*

Sterilization   A surgical procedure to make a person permanently infertile, for example, tubal ligation and vasectomy.

Subclavian arteries   The blood vessels that deliver blood to the arms.

Superior vena cava   The large vein that collects the blood of the head and neck and returns it to the right atrium.

Surrogate motherhood A technique in which the uterus of a woman who is not the natural mother is used to develop a baby; this would be employed with the "test-tube baby" technique.

Syphilis A venereal disease induced by the spirochete *Treponema pallidum*.

T

Teratogenic agent An agent that has the capacity to induce birth defects.

Testis (plural: testes; also called testicle and testicles) The male reproductive gland in which sperm and male reproductive hormones are manufactured. They are located in the scrotum.

Testosterone The major androgen, or male sex hormone.

Test-tube baby One that develops from an egg that is fertilized outside the body (in a test tube or other piece of laboratory glassware) and then implanted into a uterus by injection through the cervix.

Thalidomide A sedative that in 1960 caused an epidemic in Europe of babies born with missing and deformed limbs.

Thoracic Adjective referring to parts of the thorax (chest), for example, thoracic vertebrae.

Tissue A group of cells similar in structure and function. Tissues are often arranged into organs.

Transsexual An individual who has the body of one sex but feels that he or she is really of the opposite sex.

Trichomoniasis A vaginitis caused by infection with a protozoan parasite.

Trimester A three-month period of pregnancy, of which there are three (the first, second, and third trimesters).

Trisomy A condition in which there are three identical chromosomes instead of the usual pair in every cell of the individual.

Trophic hormone A hormone synthesized by the pituitary gland whose chief function is to stimulate another endocrine gland.

Trophoblast The thin layer of cells around the blastocyst, which eventually develops into the placenta.

Tubal ligation A surgical procedure for sterilizing a woman by tying off and severing the oviducts.

Tumor An abnormal mass of tissue, a growth. A benign tumor is one that has no tendency to spread into other tissues. A malignant tumor, a cancer, tends to spread into other parts of the body.

U

Ulcer An open sore on the skin or on a mucous membrane that tends to persist.

Umbilical cord The organ that contains the blood vessels interconnecting the embryo and fetus with the placenta; it contains two umbilical arteries and a single umbilical vein.

Urethra The tube that carries urine from the bladder to the outside of the body. It is located in the penis and is approximately 4 inches long in the man. It is much shorter in the woman and extends to the vestibule.

Urinary meatus The opening of the urethra; it is located in the vestibule in women and in the tip of the glans penis in men.

Urogenital diaphragm A triangular membrane about ¼ inch thick that fills in the pubic arch of the pelvis. The urethra and, in the woman, the vagina pass through this membrane.

Urogenital folds The external margins of the urogenital sinus, which develop into the labia minora in the female.

Urogenital sinus A slitlike open space in the base of the genital tubercle.

Uterus Also referred to as the womb. This is the pear-sized organ within which the embryo and fetus develop.

V

Vacuum aspiration Removal of the contents of the uterus (e.g., an early pregnancy) by gentle suction.

Vagina The tubelike external genital organ of the woman that receives the penis during intercourse and serves as the birth canal during parturition.

Vaginitis An inflammation of the lining of the vagina, induced by infection or trauma.

Varicocele A varicosity of the spermatic vein.

Vas deferens (plural: vasa deferentia) The duct that carries sperm from the epididymis, through the spermatic cord, inguinal canal, and pelvis, and finally to the prostatic portion of the urethra.

Vasectomy A surgical procedure for sterilizing a man by severing and tying off both vasa deferentia.

Vein A blood vessel that picks up blood from an organ and returns it to the heart.

Venereal diseases Those that are primarily or exclusively transmitted by intimate sexual contact.

Venereal warts Benign growths on the skin of the genital area produced by a virus; this is an STD.

Ventral The front side of the body (face or belly side).

Ventricles The muscular portions of the heart that receive blood from the atria and pump it to either the lungs or the rest of the body.

Vestibule The region between the labia minora containing the vaginal and urethral orifices.

Villi Finger-like projections of any biological surface that increase surface area for the purpose of interchange of fluids and dissolved substances.

Visceral arches The paired bands of tissue along the side of the embryonic head and neck that give rise to parts of the face.

Visceral grooves Depressions in the ectoderm (skin) of the sides of the face and neck that delineate the visceral arches.

Visceral muscle See *Smooth muscle.*

Vulva The region between the labia majora of women.

X

X chromosome The rather large sex chromosome present in paired form (XX) in the cells of women, but only singly in men.

Y

Y chromosome The small sex chromosome found in single (unpaired) form only in the cells of male individuals.

Yolk sac A hollow ball that forms during the second week of development and remains attached to the embryo and fetus for several months. It is a site for the development of red blood cells.

Z

Zygote The freshly fertilized egg.

Selected
Bibliography

UNIT ONE

Austin, C. R., and R. V. Short (eds.): *Reproduction in Mammals,* Book 3: *Hormones in Reproduction.* Cambridge University Press, London, 1972, 148 pp. Paperback.

Federation of Feminist Woman's Health Centers: *A New View of a Woman's Body.* Simon and Schuster, New York, 1981, 174 pp. Paperback. Different and interesting.

Hafez, E. S. E., and T. N. Evans (eds.): *Human Reproduction, Conception, and Contraception.* Harper & Row, Hagerstown, Md., 1973, 778 pp. An advanced treatise.

Hart, G.: *Sexually Transmitted Diseases.* Carolina Biology Readers, No. 95. Carolina Biological Supply Company, Burlington, N. C., 1977, 16 pp.

Hogarth, P. J.: *Biology of Reproduction.* John Wiley and Sons, New York, 1978, 189 pp. Moderately difficult.

Katchadourian, H.: *The Biology of Adolescence.* W. H. Freeman and Co., San Francisco, 1977, 274 pp. Paperback.

Lein, A.: *The Cycling Female.* W. H. Freeman and Co., San Francisco, 1979, 135 pp. Paperback. Elementary.

Madaras, L., and J. Patterson: *Womancare: A Gynecological Guide to Your Body.* Avon Paperback, New York, 1981, 938 pp.

Mader, S. S.: *Human Reproductive Biology.* Wm. C. Brown Co., Dubuque, Iowa, 1980, 276 pp. Paperback. Elementary.

Masters, W. H., and V. E. Johnson: *Human Sexual Response.* Little, Brown and Co., Boston, 1966, 366 pp.

Masters, W. H., and V. E. Johnson: *Human Sexual Inadequacy.* Little, Brown and Co., Boston, 1970, 467 pp.

Netter, F. H.: *The CIBA Collection of Medical Illustrations,* Vol. 2: *Reproductive System.* CIBA Co., Summit, N.J., 1965, 287 pp. Excellent illustrations of normal and abnormal anatomy.

Page, E. W., C. Villee, and D. Villee: *Human Reproduction: The Core Content of Obstetrics, Gynecology and Perinatal Medicine.* W. B. Saunders Co., Philadelphia, 1976, 471 pp.

There are numerous texts of human anatomy and physiology that are usually written for beginning college students in various paramedical curricula. Any of these would be useful supplementary reading for Unit One at a moderate level of difficulty. The following two have excellent sections on the reproductive and endocrine systems.

Hole, J. W., Jr.: *Human Anatomy and Physiology.* W. C. Brown Co., Dubuque, Iowa, 1978, 814 pp.

Tortora, G. J., and N. P. Anagnostakos: *Principles of Anatomy and Physiology,* 2nd ed. Canfield Press, San Francisco, 1978, 738 pp.

UNIT TWO

Austin, C. R., and R. V. Short (eds.): *Reproduction in Mammals,* Book 1: *Germ Cells and Fertilization.* Cambridge University Press, London, 1972, 136 pp.

Austin, C. R., and R. V. Short (eds.): *Reproduction in Mammals,* Book 2: *Embryonic and Fetal Development.* Cambridge University Press, London, 1972, 158 pp. Paperback.

Bing, E.: *Six Practical Lessons for an Easier Childbirth.* Bantam Books, New York, 1977, 138 pp. Paperback.

Ebert, J. D., and I. M. Sussex: *Interacting Systems in Development,* 2nd ed. Holt, Rinehart and Winston, New York, 1970, 338 pp. Paperback. A very readable account of development, emphasizing the experimental analysis of mechanisms.

Ewy, D., and R. Ewy: *Preparation for Childbirth, A Lamaze Guide.* New American Library, New York, 1976, 224 pp. Paperback.

Leboyer, F.: *Birth Without Violence.* Alfred A. Knopf, New York, 1975.

Moore, K. L.: *Before We Are Born: Basic Embryology and Birth Defects.* W. B. Saunders Co., Philadelphia, 1974, 245 pp. Paperback. A slightly abbreviated version of the following entry, written primarily for students in the allied health sciences.

Moore, K. L.: *The Developing Human: Clinically Oriented Embryology,* 2nd ed. W. B. Saunders Co., Philadelphia, 1977, 411 pp. A widely used medical text.

Nilsson, L.: *A Child Is Born.* Delacorte Press, New York, 1977, 160 pp. Beautiful photographs.

The Womanly Art of Breastfeeding. La Leche League International, Franklin Park, Ill., 1981, 368 pp. Paperback.

UNIT THREE

Apgar, V., and J. Beck: *Is My Baby All Right?* Simon and Schuster (Pocket Books), New York, 1974, 542 pp. Paperback. A detailed popular account of various birth defects.

Austin, C. R., and R. V. Short (eds.): *Reproduction in Mammals,* Book 5: *Artificial Control of Reproduction.* Cambridge University Press, London, 1972, 157 pp. Paperback.

Edwards, R. G.: *Test-tube Babies.* Carolina Biology Readers, No. 89. Carolina Biological Supply Company, Burlington, N. C., 1981, 16 pp.

Hatcher, R. A., G. K. Stewart, F. Stewart, F. Guest, P. Stratton, and A. H. Wright: *Contraceptive Technology,* 9th ed. Halsted Press, New York, 1978–1979, 192 pp. Paperback.

Kachadourian, H. A., and D. T. Lunde: *Fundamentals of Human Sexuality,* 3rd ed. Holt, Rinehart and Winston, New York, 1980, 608 pp.

McCary, J. L.: *McCary's Human Sexuality,* 3rd ed. D. van Nostrand Co., New York, 1978, 500 pp.

McKusick, V. A.: *Human Genetics.* Prentice-Hall, Englewood Cliffs, N.J., 1969, 221 pp. Paperback. A small but comprehensive book of moderate difficulty.

Saxen, L., and J. Rapola: *Congenital Defects.* Holt, Rinehart and Winston, New York, 1969, 247 pp. An excellent book of moderate difficulty.

Singer, S.: *Human Genetics.* W. H. Freeman Co., San Francisco, 1978, 139 pp. Paperback. A very readable book for those with limited biological background.

Wilson, J. G.: *Environment and Birth Defects.* Academic Press, New York, 1973, 305 pp.

Wolstenholme, G. E. W., and D. W. Fitzsimmons (eds.): *Law and Ethics of AID and Embryo Transfer.* CIBA Foundation Symposium No. 17. Associated Scientific Publishers, Amsterdam, 1973, 110 pp.

Index